Congressional Research Service

Carbon Capture: A Technology Assessment

Peter Folger, Coordinator
Specialist in Energy and Natural Resources Policy

November 5, 2013

Congressional Research Service

7-5700

www.crs.gov

R41325

CRS Report for Congress ───────────────
Prepared for Members and Committees of Congress

Summary

Carbon capture and sequestration (or carbon capture and storage, CCS) is widely seen as a critical strategy for limiting atmospheric emissions of carbon dioxide (CO_2)—the principal "greenhouse gas" linked to global climate change—from power plants and other large industrial sources. This report focuses on the first component of a CCS system, the CO_2 capture process. Unlike the other two components of CCS, transportation and geologic storage, the CO_2 capture component of CCS is heavily technology-dependent. For CCS to succeed at reducing CO_2 emissions from a significant fraction of large sources in the United States, CO_2 capture technologies would need to be deployed widely. Widespread commercial deployment would likely depend, in part, on the cost of the technology deployed to capture CO_2. This report assesses prospects for improved, lower-cost technologies for each of the three current approaches to CO_2 capture: post-combustion capture; pre-combustion capture; and oxy-combustion capture.

While all three approaches are capable of high CO_2 capture efficiencies (typically about 90%), the major drawbacks of current processes are their high cost and the large energy requirements for operation. Another drawback in terms of their availability for greenhouse gas mitigation is that at present, there are still no full-scale applications of CO_2 capture on a coal-fired or gas-fired power plant (i.e., a scale of several hundred megawatts of plant capacity). To address the current lack of demonstrated capabilities for full-scale CO_2 capture at power plants, a number of large-scale demonstration projects at both coal combustion and gasification-based power plants are planned or underway in the United States and elsewhere. Substantial research and development (R&D) activities are also underway in the United States and elsewhere to develop and commercialize lower-cost capture systems with smaller energy penalties. Current R&D activities include development and testing of new or improved solvents that can lower the cost of current post-combustion and pre-combustion capture, as well as research on a variety of potential "breakthrough technologies" such as novel solvents, sorbents, membranes, and oxyfuel systems that hold promise for even lower-cost capture systems.

In general, the focus of most current R&D activities is on cost reduction rather than additional gains in the efficiency of CO_2 capture (which can result in cost increases rather than decreases). Key questions regarding the outcomes from these R&D efforts are when advanced CO_2 capture systems would be available for commercial rollout, and how much cheaper they would be compared to current technology. "Technology roadmaps" developed by governmental and private-sector organizations in the United States and elsewhere anticipate that CO_2 capture will be available for commercial deployment at power plants by 2020. A number of roadmaps also project that some novel, lower-cost technologies would be commercial in the 2020 time frame. Such projections acknowledge, however, that this would require aggressive and sustained efforts to advance promising concepts to commercial reality.

Achieving significant cost reductions would likely require not only a vigorous and sustained level of R&D, but also a significant market for CO_2 capture technologies to generate a substantial level of commercial deployment. At present such a market does not exist. While various types of incentive programs can accelerate the development and deployment of CO_2 capture technology, actions that significantly limit emissions of CO_2 to the atmosphere ultimately would be needed to realize substantial and sustained reductions in the future cost of CO_2 capture.

Contents

Figures

Tables

Contacts

Introduction

Congressional interest has grown in carbon capture and sequestration (or carbon capture and storage, CCS) as part of legislative strategies to mitigate global climate change. The promise of CCS lies in the potential for technology to capture CO_2 emitted from large, industrial sources, thus significantly decreasing CO_2 emissions without drastically changing U.S. dependence on fossil fuels, particularly coal, for electricity generation. The future use of coal—a significant component of the U.S. energy portfolio—in the United States would likely depend on whether and how CCS is deployed if legislative or regulatory actions curtail future CO_2 emissions.

Unlike the other two components of CCS, transportation and geologic storage, the first component of CCS—CO_2 capture—is almost entirely technology-dependent. For CCS to succeed at reducing CO_2 emissions from a significant fraction of large sources in the United States, CO_2 capture technology would need to deployed widely. Widespread commercial deployment would likely depend on the cost of capturing CO_2. This report examines the factors underlying the cost of currently available CO_2 capture technologies and advanced capture systems. This report also examines efforts to commercialize other advanced technologies, namely sulfur dioxide (SO_2) and nitrogen oxide (NOx) capture technologies to reduce air pollution, to glean insights that could be useful for assessing the prospects for improved, lower-cost CO_2 capture systems.

The transportation and storage components of CCS are not nearly as technology-dependent as the capture component. Nonetheless, transportation and sequestration costs, while generally much smaller than capture costs, could be very high in some cases. They would depend, in part, on how long it would take to reach an agreement on a regulatory framework to guide long-term CO_2 injection and storage, and on what those regulations would require. CCS deployment would also depend on the degree of public acceptance of a large-scale CCS enterprise. This report provides a "snapshot" of current technological development, but is both prospective and retrospective in that it also examines emerging or advanced technologies that may affect future CCS deployment, and looks at lessons from past experience with large-scale technological development and deployment as guidelines that could be used to shape energy policy.

Authorship and Structure of the Report

This technology assessment and report was undertaken by Carnegie Mellon University, Department of Engineering and Public Policy, under the leadership of Edward S. Rubin, together with Aaron Marks, Hari Mantripragada, Peter Versteeg, and John Kitchin. The work was performed under contract to CRS, and is part of a multiyear CRS project to examine different aspects of U.S. energy policy. Peter Folger, CRS Specialist in Energy and Natural Resources Policy, served as the CRS project coordinator.

The bulk of the report consists of 10 chapters, together with figures and tables. Each chapter can be read independently; however, "Chapter 1: Executive Summary," "Chapter 2: Background and Scope of Report," and "Chapter 3: Overview of CO_2 Capture Technologies" provide the reader with background and context for a more complete understanding of some of the more technologically focused discussions in other chapters.

The material in this report is current as of July 19, 2010. The report will not be updated.

Acknowledgment

This report was funded, in part, by a grant from the Joyce Foundation.

Chapter 1: Executive Summary

Background

Carbon capture and storage (CCS) is widely seen as a critical technology for limiting atmospheric emissions of carbon dioxide (CO_2)—the principal "greenhouse gas" linked to global climate change—from power plants and other large industrial sources. This report focuses on the first component of a CCS system, namely, the CO_2 capture process. The goal of the report is to provide a realistic assessment of prospects for improved, lower-cost technologies for each of the three current approaches to CO_2 capture, namely, post-combustion capture from power plant flue gases using amine-based solvents such as monoethanolamine (MEA) and ammonia; pre-combustion capture (also via chemical solvents) from the synthesis gas produced in an integrated coal gasification combined cycle (IGCC) power plant; and oxy-combustion capture, in which high-purity oxygen rather than air is used for combustion in a pulverized coal (PC) power plant to produce a flue gas with a high concentration of CO_2 amenable to capture without a post-combustion chemical process.

Currently, post-combustion and pre-combustion capture technologies are commercial and widely used for gas stream purification in a variety of industrial processes. Several small-scale installations also capture CO_2 from power plant flue gases to produce CO_2 for sale as an industrial commodity. Oxy-combustion capture, however, is still under development and is not currently commercial.

The advantages and limitations of each of these three methods are discussed in this report, along with plans for their continued development. While all three approaches are capable of high CO_2 capture efficiencies (typically about 90%), the major drawbacks of current processes are their high cost and the large energy requirement for operation (which significantly reduces the net plant capacity and contributes to the high cost of capture). Another drawback in terms of their availability for greenhouse gas mitigation is that at present, there are still no applications of CO_2 capture on a coal-fired or gas-fired power plant at full scale (i.e., a scale of several hundred megawatts of plant capacity).

Current Research and Development (R&D) Activities

To address the current lack of demonstrated capabilities for full-scale CO_2 capture at power plants, a number of large-scale demonstration projects at both coal combustion and gasification-based power plants are planned or underway in the United States and elsewhere. The current status of these projects and the technologies they plan to employ are summarized in the body of this report. Most of these demonstrations are expected to begin operation in 2014 or 2015. Planned projects for other types of industrial facilities also are discussed.

Also elaborated in this report are the substantial R&D activities underway in the United States and elsewhere to develop and commercialize lower-cost capture systems with smaller energy penalties. To characterize the status of capture technologies and the prospects for their commercial availability, five stages of development are defined in this report: conceptual designs; laboratory or bench scale; pilot plant scale; full-scale demonstration plants; and commercial processes. Current activities at each of these stages are reviewed for each of the three major capture routes.

Current R&D activities include development and testing of new or improved solvents that can lower the cost of current post-combustion and pre-combustion capture, as well as research on a variety of potential "breakthrough technologies" such as novel solvents, sorbents, membranes, and oxyfuel systems that hold promise for even lower-cost capture systems. Most of the latter processes, however, are still in the early stages of research and development (i.e., conceptual designs and laboratory- or bench-scale processes), so that credible estimates of their performance and (especially) cost are lacking at this time. **Table 1** lists the major approaches being pursued for post-combustion capture, although many of these approaches apply to pre-combustion and oxy-combustion capture as well.

**Table 1. Post-Combustion Capture Approaches Being Developed
at Laboratory or Bench Scale**

Liquid Solvents	Solid Adsorbents	Membranes
Advanced amines	Supported amines	Polymeric
Potassium carbonate	Carbon-based	Amine-doped
Advanced mixtures	Sodium carbonate	Integrated with absorption
Ionic liquids	Crystalline materials	Biomimetic-based

Source: Edward S. Rubin, Aaron Marks, Hari Mantripragada, Peter Versteeg, and John Kitchin, Carnegie Mellon University, Department of Engineering and Public Policy.

Processes under development at the more advanced pilot plant scale are, for the most part, new or improved solvent formulations (such as ammonia and advanced amines) that are undergoing testing and evaluation. These advanced solvents could be available for commercial use within several years if subsequent full-scale testing confirms their overall benefit. Pilot-scale oxy-combustion processes also are currently being tested and evaluated for planned scale-up, while two IGCC power plants in Europe are installing pilot plants to evaluate pre-combustion capture options.

In general, the focus of most current R&D activities is on cost reduction rather than additional gains in the efficiency of CO_2 capture (which can result in cost increases rather than decreases). A number of R&D programs emphasize the need for lower-cost retrofit technologies suitable for existing power plants. As a practical matter, however, most technologies being pursued to reduce capture costs for new plants also apply to existing plants. Indeed, as the fleet of existing coal-fired power plants continues to age, the size of the potential U.S. retrofit market for CO_2 capture will continue to shrink, as older plants may not be economic to retrofit (although the situation in other countries, especially China, may be quite different).

Future Outlook

Whether for new power plants or existing ones, the key questions are the same: When will advanced CO_2 capture systems be available for commercial rollout, and how much cheaper will they be compared to current technology?

To address the first question, this report reviews a variety of "technology roadmaps" developed by governmental and private-sector organizations in the United States and elsewhere. All of these roadmaps anticipate that CO_2 capture will be available for commercial deployment at power plants by 2020. Current commercial technologies like post-combustion amine systems could be available sooner. A number of roadmaps also project that novel, lower-cost technologies like solid sorbent systems for post-combustion capture will be commercial in the 2020 time frame. Such projections acknowledge, however, that this will require aggressive and sustained efforts to advance promising concepts to commercial reality.

That caveat is strongly supported by a review of experience from other recent R&D programs to develop lower-cost technologies for post-combustion SO_2 and NO_x capture at coal-fired power plants. Those efforts typically took two decades or more to bring new concepts (like combined SO_2 and NO_x capture processes) to commercial availability. By then, however, the cost advantages initially foreseen for these novel systems had largely evaporated in most cases: the advanced technologies tended to get more expensive as their development progressed (consistent with "textbook" descriptions of the innovation process), while the cost of formerly "high-cost" commercial technologies gradually declined over time. The absence of a significant market for the novel technologies put them at a further disadvantage. This is similar to the situation for CO_2 capture systems today. Thus, the development of advanced CO_2 capture technologies is not without risks.

With regard to future cost reductions, the good news based on past experience is that the costs of environmental technologies that succeed in the marketplace tend to fall over time. For example, after an initial rise during the early commercialization period, the cost of post-combustion SO_2 and NO_x capture systems declined by 50% or more after about two decades of deployment at coal-fired power plants. This trend is consistent with the "learning curve" behavior seen for many other classes of technology. It thus appears reasonable to expect a similar trend for future CO_2 capture costs once these technologies become widely deployed. Note, too, that the cost of CO_2 capture also depends on other aspects of power plant design, financing, and operation—not solely on the cost of the CO_2 capture unit. Future improvements in net power plant efficiency, for example, will tend to lower the unit cost of CO_2 capture.

Other cost estimates for advanced CO_2 capture systems are based on engineering-economic analysis of proposed system designs. For example, recent studies by the U.S. Department of Energy (DOE) foresee the cost of advanced PC and IGCC power plants with CO_2 capture falling by 27% and 31%, respectively, relative to current costs as a result of successful R&D programs. No estimates are provided, however, as to when the various improvements described are expected be commercially available. In general, however, the farther away a technology is from commercial reality, the lower its estimated cost tends to be. Thus, there is considerable uncertainty in cost estimates for technologies that are not yet commercial, especially those that exist only as conceptual designs.

More reliable estimates of future technology costs typically are linked to projections of their expected level of commercial deployment in a given time frame (i.e., a measure of their market size). For power plant technologies like CO_2 capture systems, this is commonly expressed as total installed capacity. However, as with other technologies whose sole purpose is to reduce environmental emissions, there is no significant market for power plant CO_2 capture systems absent government actions or policies that effectively create such markets—either through regulations that limit CO_2 emissions, or through voluntary incentives such as tax credits or direct financial subsidies. The technical literature and historical evidence examined in this report strongly link future cost reductions for CO_2 capture systems to their level of commercial deployment. In widely used models based on empirical "experience curves," the latter measure serves as a surrogate for the many factors that influence future technology costs, including the level of R&D expenditures and the new knowledge gained through learning-by-doing (related to manufacturing) and learning-by-using (related to technology use).

Based on such models, published estimates project the future cost of electricity from power plants with CO_2 capture to fall by as much as 30% below current values after roughly 100,000 megawatts (MW) of capture plant capacity is installed and operated worldwide. That estimate is in line with the DOE projects noted above. If achieved, it would represent a significant decrease from current costs—one that would bring the cost and efficiency of future power plants with CO_2 capture close to that of current plants without capture. For reference, it took approximately 20 years following passage of the 1970 Clean Air Act Amendments to achieve a comparable level of technology deployment for SO_2 capture systems at coal-fired power plants.

Uncertainty estimates for these projections, however, indicate that future cost reductions for CO_2 capture also could be much smaller than indicated above. Thus, whether future cost reductions will meet, exceed, or fall short of current estimates will only be known with hindsight.

In the context of this report, the key insight governing prospects for improved carbon capture technology is that achieving significant cost reductions will require not only a vigorous and sustained level of R&D, but also a substantial level of commercial deployment. That will necessitate a significant market for CO_2 capture technologies, which can only be established by government actions. At present such a market does not yet exist. While various types of incentive programs can accelerate the development and deployment of CO_2 capture technology, actions that significantly limit emissions of CO_2 to the atmosphere ultimately are needed to realize substantial and sustained reductions in the future cost of CO_2 capture.

Chapter 2: Background and Scope of Report

Introduction

Global climate change is an issue of major international concern and the focus of proposed mitigation policy measures in the United States and elsewhere. In this context, the technology of carbon capture and storage (CCS) has received increasing attention over the past decade as a potential method of limiting atmospheric emissions of carbon dioxide (CO_2)—the principal "greenhouse gas" linked to climate change.

Worldwide interest in CCS stems principally from three factors. First is a growing consensus that large reductions in global CO_2 emissions are needed to avoid serious climate change impacts.[1] Because electric power plants are a major source of GHG emissions, their emissions must be significantly curtailed.

Second is the realization that large emission reductions cannot be achieved easily or quickly simply by using less energy or by replacing fossil fuels with alternative energy sources that emit little or no CO_2. The reality is that the world (and the United States itself) today relies on fossil fuels for over 85% of its energy use. Changing that picture dramatically will take time. CCS thus offers a way to get large CO_2 reductions from power plants and other industrial sources until cleaner, sustainable technologies can be widely deployed.

Finally, energy-economic models show that adding CCS to the suite of other GHG reduction measures significantly lowers the cost of mitigating climate change. Studies also have affirmed that by 2030 and beyond, CCS is a major component of a cost-effective portfolio of emission reduction strategies.[2]

Figure 1 depicts the overall CCS process applied to a power plant or other industrial process. The CO_2 produced from carbon in the fossil fuels or biomass feedstock is first captured, then compressed to a dense liquid to facilitate its transport and storage. The main storage option is underground injection into a suitable geological formation.

At the present time, CCS is not yet commercially proven in the primary large-scale application for which it is envisioned—electric power plants fueled by coal or natural gas. Furthermore, the cost of CCS today is relatively high, due mainly to the high cost of CO_2 capture (which includes the cost of CO_2 compression needed for transport and storage). This has prompted a variety of governmental and private-sector research programs in the United States and elsewhere to develop more cost-effective methods of CO_2 capture.

---

[1] National Research Council, *America's Climate Choices: Limiting the Magnitude of Future Climate Change*, The National Academies Press, Washington, DC, May 2010; S. Solomon et al., eds., *Climate Change 2007: The Physical Science Basis*, Contribution of Working Group I to the Fourth Assessment Report of the Intergovernmental Panel on Climate Change. Cambridge University Press, Cambridge, UK and New York, NY, 2007.

[2] J. Edmonds, "The Potential Role of CCS in Climate Stabilization," Proc. *9th International Conference on Greenhouse Gas Control Technologies, 2008,* Washington, DC; B. Metz, et al., eds., *Climate Change 2007: Mitigation. Contribution of Working Group III to the Fourth Assessment Report of the Intergovernmental Panel on Climate Change* Cambridge University Press, Cambridge, United Kingdom and New York, NY, USA.

Figure 1. Schematic of a CCS System, Consisting of CO₂ Capture, Transport, and Storage

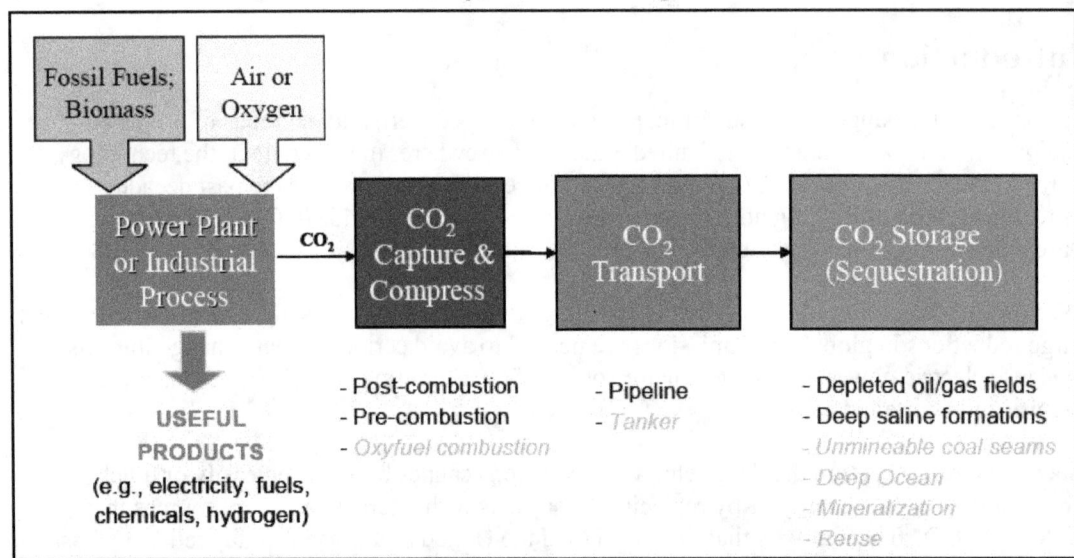

Source: E. S. Rubin, "Will Carbon Capture and Storage be Available in Time?," American Association for the Advancement of Science, Annual Meeting, San Diego, CA, February 18-22, 2010.

Notes: Carbon inputs may include fossil fuels and biomass. Technical options are listed below each stage. Those in italics are not yet available or implemented at a commercial scale.

Report Objectives and Scope

The present report seeks to assist the Congressional Research Service (CRS) in providing analysis and information to the U.S. Congress related to national policy on climate change. More specifically, the objective is to provide a realistic assessment of prospects for improved, lower-cost CO_2 capture systems for use at power plants and in other industrial processes. Issues and technologies associated with CO_2 transport and storage are thus outside the scope of this report. The tasks in the statement of work for this study were to:

- Discuss the advantages, as well as the possible limitations, on continued development and commercial deployment of each of the three current approaches to CO_2 capture, namely (1) post-combustion chemical treatment and capture of flue gas CO_2 with amines, such as monoethanolamine (MEA) and ammonia; (2) pre-combustion chemical removal of CO_2 from the synthesis gas produced from coal in an integrated gasification combined cycle (IGCC) plant; and (3) oxyfuel combustion, in which pure oxygen replaces the air normally used in coal combustion to produce a flue gas containing mainly water vapor and concentrated CO_2, which is amenable to capture without a post-combustion chemical process.

- Investigate research in the United States and elsewhere to assess (1) the evolution of current technologies, especially whether significant gains in the efficiency of CO_2 capture, and thus cost reductions, can be reasonably expected for the technologies discussed above, along with reasonable estimates of the commercial rollout schedules for retrofit and new plant use; and (2) the potential of emerging and "breakthrough technologies" such as advanced catalysts for CO_2 conversion, novel solvents, sorbents, membranes, and thin films for gas separation. This part of the study describes where such technologies currently are in the R&D process (e.g., concept, laboratory, pilot scale and so on), in order to provide Congress with an understanding of whether the research focus is on engineering and technology development of new processes whose physics and chemistry are well understood, as distinguished from projects whose research focus is on first principles and conceptual design, with the engineering of an actual device still many years in the future.

Organization of This Report

Consistent with the above objectives, this report's "Chapter 3: Overview of CO_2 Capture Technologies" first gives an overview of CO_2 capture technologies and their application to new and existing facilities. The current costs of CO_2 capture also are presented. "Chapter 4: Stages of Technology Development" then discusses the process of technological change and defines the five stages of technological development used in this report to describe the status of CO_2 capture technologies. "Chapter 5: Status of Post-Combustion Capture," "Chapter 6: Status of Pre-Combustion Capture," and "Chapter 7: Status of Oxy-Combustion Capture" elaborate on each of the three major categories of CO_2 capture systems, namely, post-combustion, pre-combustion, and oxy-combustion capture, respectively. For each category, the current status of technology in each stage of development is described along with the technical challenges that must be overcome to move forward. "Chapter 8: Cost and Deployment Outlook for Advanced Capture Systems" then discusses the prospects for improved, lower-cost capture technologies and the timetables for commercialization projected by governmental and private-sector organizations involved in capture technology R&D. For perspective, "Chapter 9: Lessons from Past Experience" looks retrospectively at recent experience on the pace of technology innovation and deployment to control other power plant pollutants. It also discusses some of the key drivers of technology innovation that influence future prospects for carbon capture systems. Finally, "Chapter 10: Discussion and Conclusions" discusses the key findings and conclusions from this study.

Chapter 3: Overview of CO₂ Capture Technologies

Introduction

A variety of technologies for separating (capturing) CO_2 from a mixture of gases are commercially available and widely used today, typically as a purification step in an industrial process. **Figure 2** illustrates the variety of technical approaches available. The choice of technology depends on the requirements for product purity and on the conditions of the gas stream being treated (such as its temperature, pressure, and CO_2 concentration). Common applications for CO_2 capture systems include the removal of CO_2 impurities in natural gas treatment and the production of hydrogen, ammonia, and other industrial chemicals. In most cases, the captured CO_2 stream is simply vented to the atmosphere. In a few cases it is used in the manufacture of other chemicals.[3]

Figure 2. Technical Options for CO₂ Capture

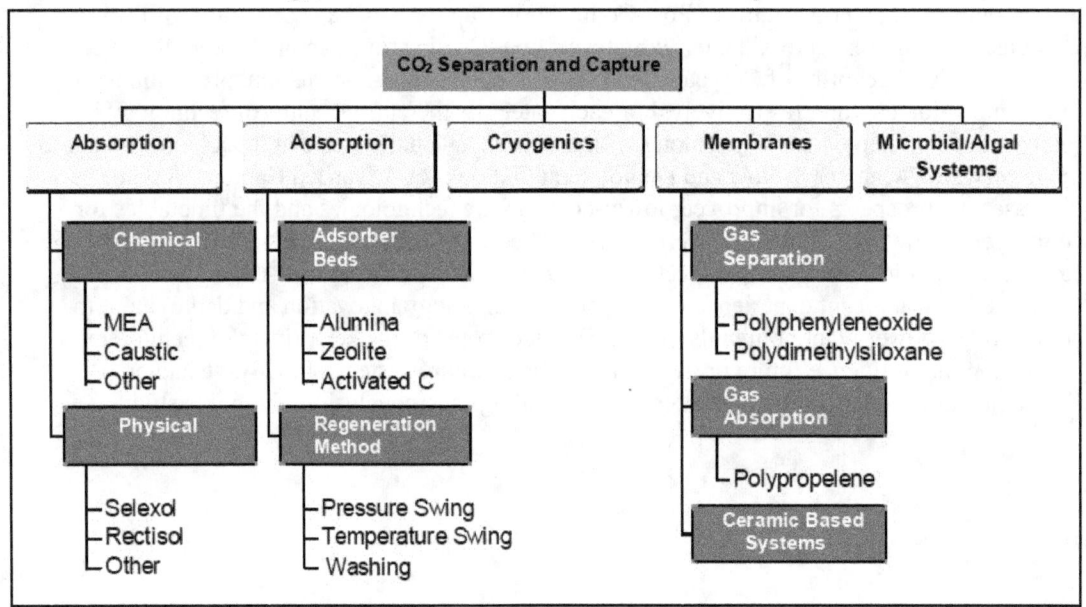

Source: A. B. Rao and E. S. Rubin, "A Technical, Economic and Environmental Assessment of Amine-Based CO₂ Capture Technology for Power Plant Greenhouse Gas Control," *Environmental Science & Technology*, vol. 36, no. 20 (2002), pp. 4467-4475.

Notes: The choice of method depends strongly on the particular application.

CO_2 also has been captured from a portion of the flue gases produced at power plants burning coal or natural gas. Here, the captured CO_2 is sold as a commodity to nearby industries such as food processing plants. Globally, however, only a small amount of CO_2 is utilized to manufacture industrial products and nearly all of it is soon emitted to the atmosphere (for example, from carbonated drinks).

---

[3] B. Metz et al., eds., *Special Report on Carbon Dioxide Capture and Storage*, Prepared by Working Group III of the Intergovernmental Panel on Climate Change. Cambridge University Press, Cambridge, UK and New York, NY, p 442, 2005.

Since most anthropogenic CO_2 is a by-product of the combustion of fossil fuels, CO_2 capture technologies, when discussed in the context of CCS, are commonly classified as either pre-combustion or post-combustion systems, depending on whether carbon (in the form of CO_2) is removed before or after a fuel is burned. A third approach, called oxyfuel or oxy-combustion, does not require a CO_2 capture device. This concept is still under development and is not yet commercial. Other industrial processes that do not involve combustion employ the same types of CO_2 capture systems that would be employed at power plants.

In all cases, the aim is to produce a stream of pure CO_2 that can be permanently stored or sequestered, typically in a geological formation. This requires high pressures to inject CO_2 deep underground. Thus, captured CO_2 is first compressed to a dense "supercritical" state, where it behaves as a liquid that can be readily transported via pipeline and injected into a suitable geological formation. However, the CO_2 compression step is commonly included as part of the capture system, since it is usually located at the industrial plant site where CO_2 is captured.

Post-Combustion Processes

As the name implies, these systems capture CO_2 from the flue gases produced after fossil fuels or other carbonaceous materials (such as biomass) are burned. Combustion-based power plants provide most of the world's electricity today. In a modern coal-fired power plant, pulverized coal (PC) is mixed with air and burned in a furnace or boiler. The heat released by combustion generates steam, which drives a turbine-generator (**Figure 3**). The hot combustion gases exiting the boiler consist mainly of nitrogen (from air) plus smaller concentrations of water vapor and CO_2 formed from the hydrogen and carbon in the fuel. Additional products formed during combustion from impurities in coal include sulfur dioxide, nitrogen oxides, and particulate matter (fly ash). These regulated air pollutants, as well as other trace species such as mercury, must be removed to meet applicable emission standards. In some cases, additional removal of pollutants (especially SO_2) is required to provide a sufficiently clean gas stream for subsequent CO_2 capture.

Figure 3. Schematic of a Coal-Fired Power Plant with Post-Combustion CO_2 Capture Using an Amine Scrubber System

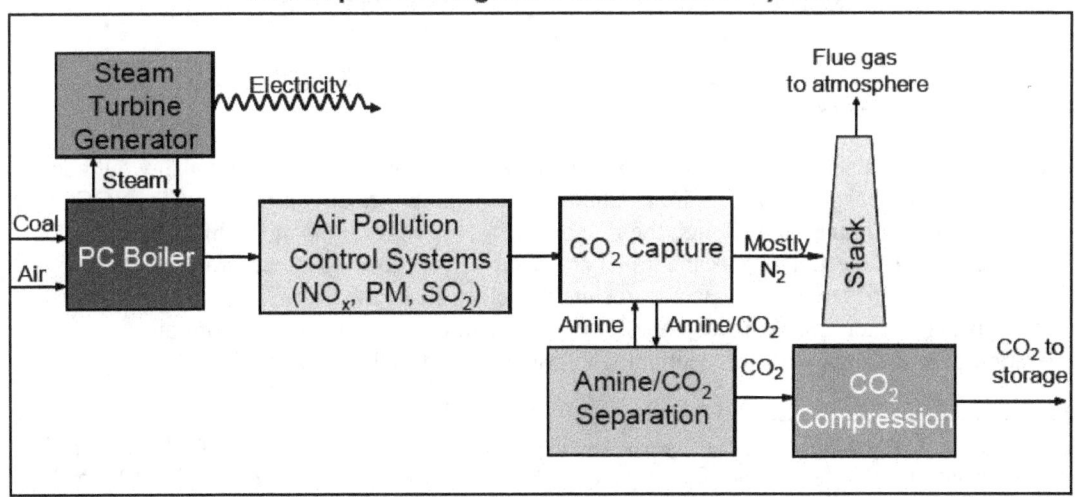

Source: E. S. Rubin, "CO_2 Capture and Transport," *Elements*, vol. 4 (2008), pp. 311-317.

Notes: Other major air pollutants (nitrogen oxides, particulate matter, and sulfur dioxide) are removed from the flue gas prior to CO_2 capture.

With current technology, the most effective method of CO_2 capture from the flue gas of a PC plant is by chemical reaction with an organic solvent such as monoethanolamine (MEA), one of a family of amine compounds. In a vessel called an absorber, the flue gas is "scrubbed" with an amine solution, typically capturing 85% to 90% of the CO_2. The CO_2-laden solvent is then pumped to a second vessel, called a regenerator, where heat is applied (in the form of steam) to release the CO_2. The resulting stream of concentrated CO_2 is then compressed and piped to a storage site, while the depleted solvent is recycled back to the absorber. **Figure 4** shows details of a post-combustion capture system design.

Figure 4. Details of Flue Gas and Sorbent Flows for an Amine-Based Post-Combustion CO_2 Capture System

(absorber is shown on the left, and regenerator on the right)

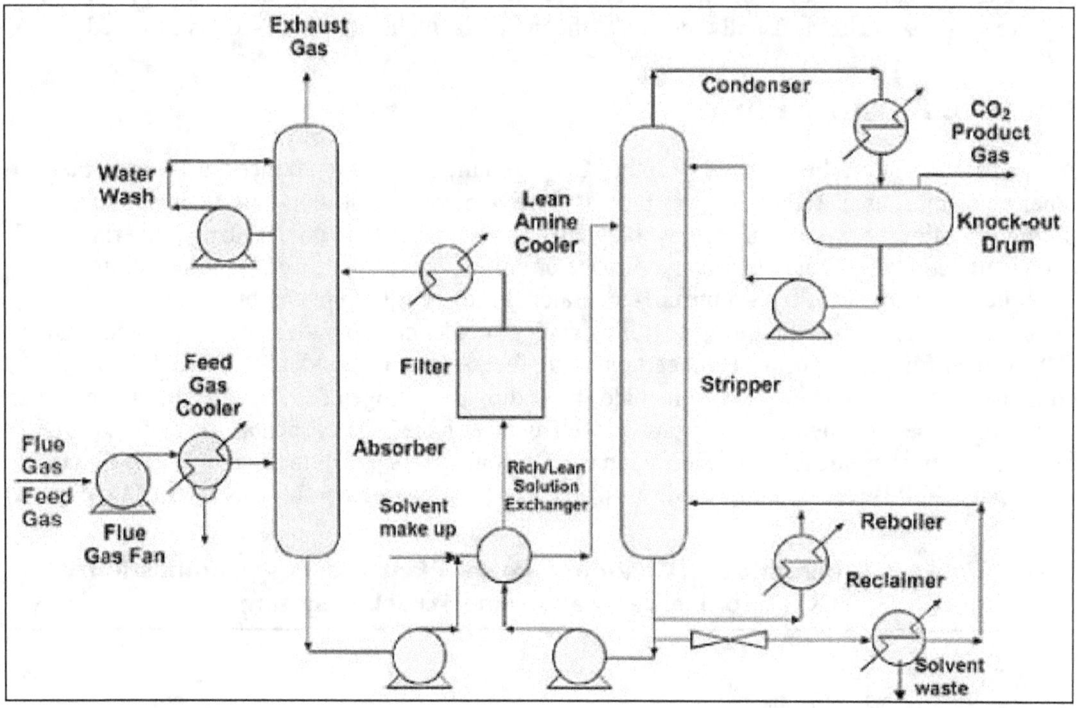

Source: Metz, *Special Report.*

The same post-combustion capture technology that would be used at a PC plant also would be used for post-combustion CO_2 capture at a natural gas-fired boiler or combined cycle (NGCC) power plant (see **Figure 5**). Although the flue gas CO_2 concentration is more dilute than in coal plants, high removal efficiencies can still be achieved with amine-based capture systems. The absence of impurities in natural gas also results in a clean flue gas stream, so that no additional cleanup is needed for effective CO_2 capture. Further details on the design, performance, and operation of amine-based capture technologies can be found in the technical literature.[4]

---

[4] A. B. Rao and E. S. Rubin, "A Technical, Economic and Environmental Assessment of Amine-Based CO2 Capture Technology for Power Plant Greenhouse Gas Control," *Environmental Science & Technology*, vol. 36 (2002), pp. 4467-4475; Metz, *Special Report*. U.S. Department of Energy (DOE), *Cost and Performance Baseline for Fossil Energy Plants. Volume 1: Bituminous Coal and Natural Gas to Electricity Final Report,* National Energy Technology Laboratory, Pittsburgh, PA, August 2007.

Figure 5. Schematic of an Amine-Based Post-Combustion CO₂ Capture System Applied to a Natural Gas Combined Cycle (NGCC) Power Plant

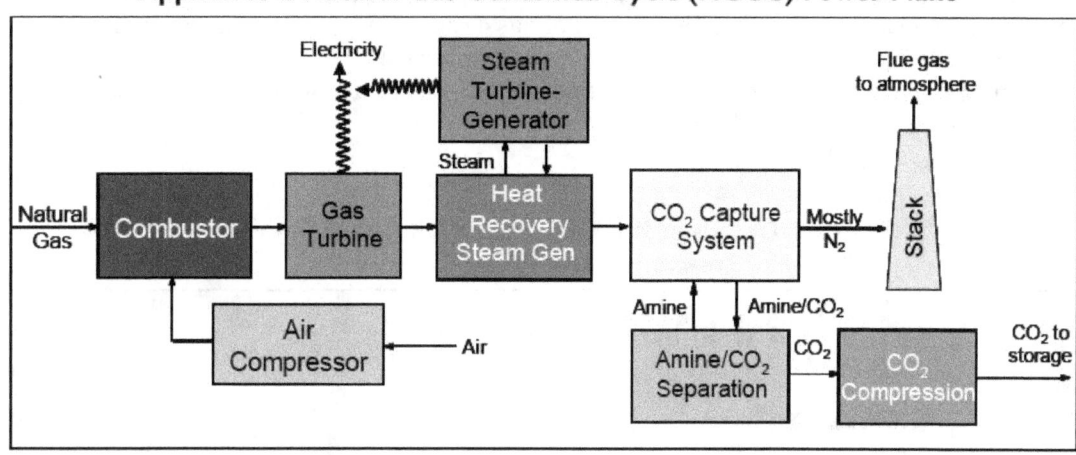

Source: Rubin, "CO₂ Capture."

Pre-Combustion Processes

To remove carbon from fuel prior to combustion, it must first be converted to a form amenable to capture. For coal-fueled plants, this is accomplished by reacting coal with steam and oxygen at high temperature and pressure, a process called partial oxidation, or gasification. The result is a gaseous fuel consisting mainly of carbon monoxide and hydrogen—a mixture known as synthesis gas, or syngas—which can be burned to generate electricity in a combined cycle power plant similar to the NGCC plant described above. This approach is known as integrated gasification combined cycle (IGCC) power generation. After particulate impurities are removed from the syngas, a two-stage "shift reactor" converts the carbon monoxide to CO_2 via a reaction with steam (H_2O). The result is a mixture of CO_2 and hydrogen. A chemical solvent, such as the widely used commercial product Selexol (which employs a glycol-based solvent), then captures the CO_2, leaving a stream of nearly pure hydrogen that is burned in a combined cycle power plant to generate electricity, as depicted in **Figure 6**.

Figure 6. Schematic of an Integrated Gasification Combined Cycle (IGCC) Coal Power Plant with Pre-Combustion CO₂ Capture Using a Water-Gas Shift Reactor and a Selexol CO₂ Separation System

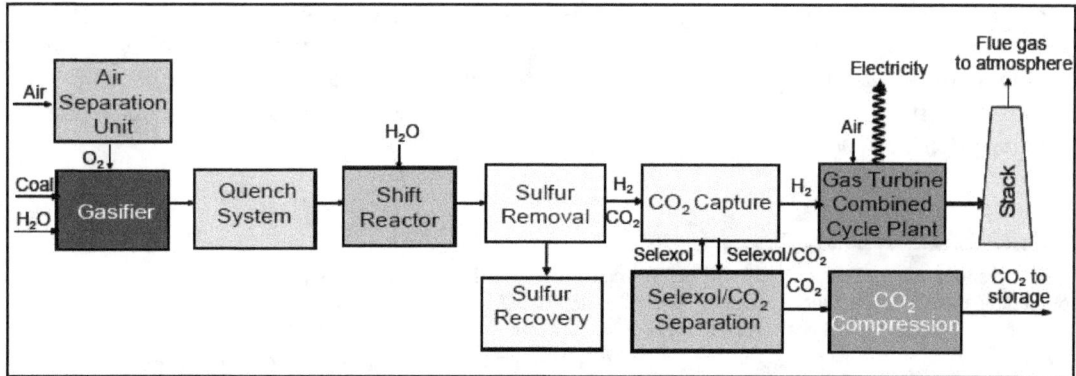

Source: Rubin, "CO₂ Capture."

Although the fuel conversion steps of an IGCC plant are more elaborate and costly than traditional coal combustion plants, CO_2 separation is much easier and cheaper because of the high operating pressure and high CO_2 concentration of this design. Thus, rather than requiring a chemical reaction to capture CO_2 (as with amine systems in post-combustion capture), the mechanism employed in pre-combustion capture involves physical adsorption onto the surface of a solvent, followed by release of the CO_2 when the sorbent pressure is dropped, typically in several stages, as depicted in **Figure 7**.

Figure 7. Details of the Fuel Gas and Sorbent Flows for Pre-Combustion CO₂ Capture

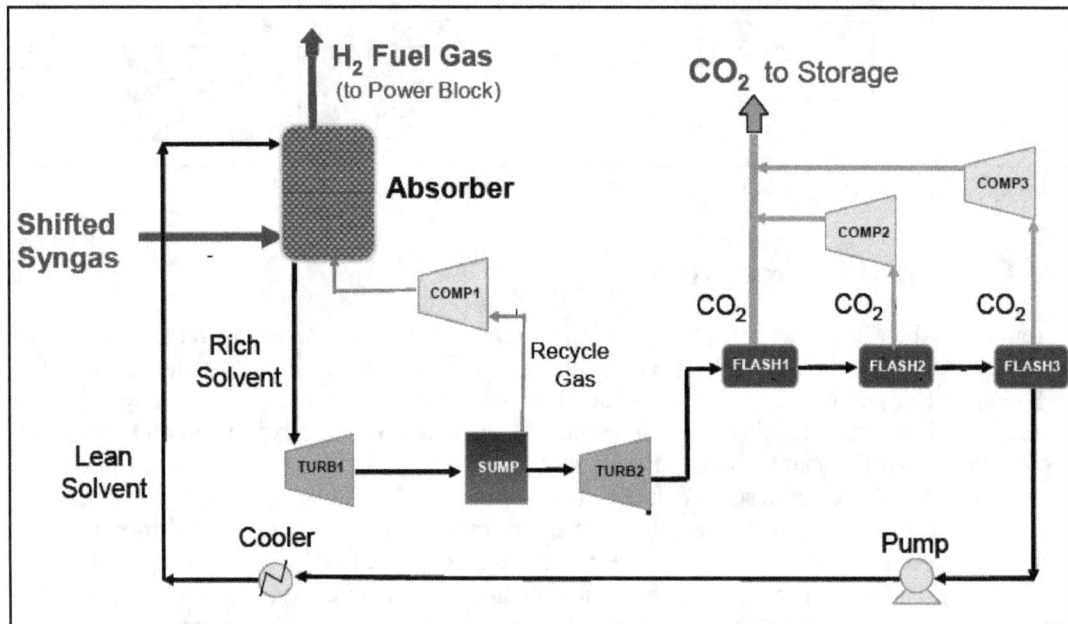

Source: Adapted from C. Chen, "A Technical and Economic Assessment of CO_2 Capture Technology for IGCC Power Plants"(Ph.D. thesis, Carnegie Mellon University, Pittsburgh, PA, 2005).

Pre-combustion capture also can be applied to power plants using natural gas. As with coal, the raw gaseous fuel is first converted to syngas via reactions with oxygen and steam—a process called reforming. This is again followed by a shift reactor and CO_2 separation, yielding streams of concentrated CO_2 (suitable for storage) and hydrogen. This is the dominant method used today to manufacture hydrogen. If the hydrogen is burned to generate electricity, as in an IGCC plant, we have pre-combustion capture. While pre-combustion CO_2 capture is usually more costly than post-combustion capture for natural gas-fired plants, some power plants of this type have been proposed.[5] Further details regarding the design, performance, and operation of pre-combustion capture systems can be found in the literature.[6]

---

[5] Scottish and Southern Energy, "SSE, BP and Partners Plan Clean Energy Plant in Scotland," at http://www.scottish-southern.co.uk/SSEInternet/index.aspx?id=894&TierSlicer1_TSMenuTargetID=444&TierSlicer1_TSMenuTargetType=1&TierSlicer1_TSMenuID=6.

[6] Metz, *Special Report.* C. Chen and E. S. Rubin, "CO_2 Control Technology Effects on IGCC Plant Performance and Cost," *Energy Policy*, vol. 37, no. 3 (2009), pp. 915-924.

Oxy-Combustion Systems

Oxy-combustion (or oxyfuel) systems are being developed as an alternative to post-combustion CO_2 capture for conventional coal-fired power plants. Here, pure oxygen rather than air is used for combustion. This eliminates the large amount of nitrogen in the flue-gas stream. After the particulate matter (fly ash) is removed, the flue gas consists only of water vapor and CO_2, plus smaller amounts of pollutants such as sulfur dioxide (SO_2) and nitrogen oxides (NO_x). The water vapor is easily removed by cooling and compressing the flue gas. Additional removal of air pollutants leaves a nearly pure CO_2 stream that can be sent directly to storage, as depicted in **Figure 8**.

Figure 8. Schematic of a Coal-Fired Power Plant Using Oxy-Combustion

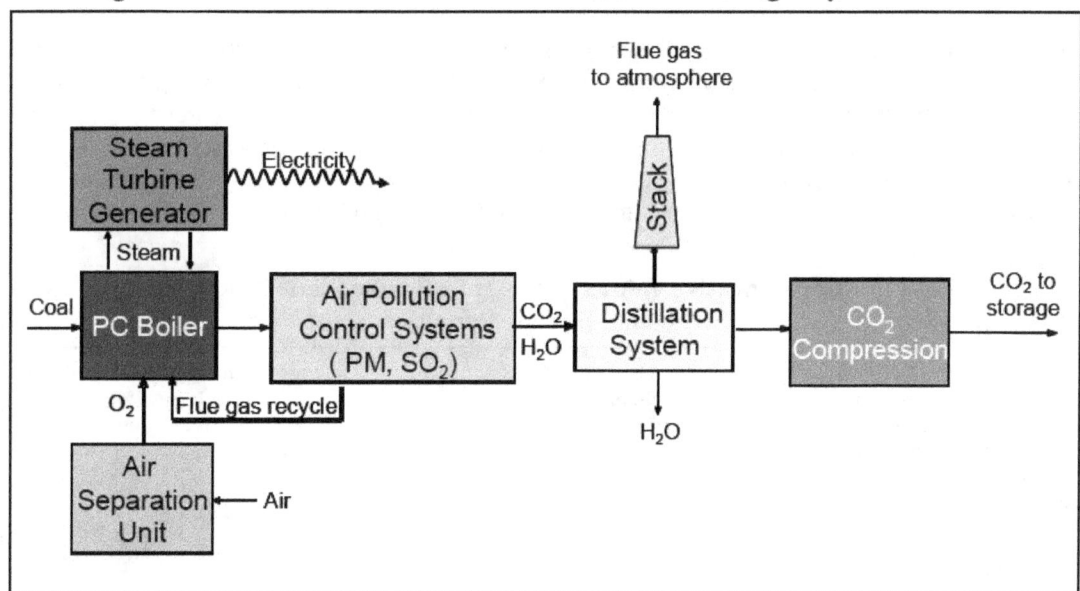

Source: Rubin, "CO₂ Capture."

The principal attraction of oxy-combustion is that it avoids the need for a costly post-combustion CO_2 capture system. Instead, however, it requires an air separation unit (ASU) to generate the relatively pure (95%-99%) oxygen needed for combustion. Roughly three times more oxygen is needed for oxyfuel systems than for an IGCC plant of comparable size, so the ASU adds significantly to the cost. Typically, additional flue gas processing also is needed to reduce the concentration of conventional air pollutants, so as to comply with applicable environmental standards, or to prevent the undesirable buildup of a substance in the flue gas recycle loop, or to achieve pipeline CO_2 purity specifications (whichever requirement is the most stringent). Because combustion temperatures with pure oxygen are much higher than with air, oxy-combustion also requires a large portion (roughly 70%) of the inert flue gas stream to be recycled back to the boiler in order to maintain normal operating temperatures. To avoid unacceptable levels of oxygen and nitrogen in the flue gas, the system also has to be carefully sealed to prevent any leakage of air into the flue gas. This is a challenge since such leakage commonly occurs at existing power plants at flanges and joints along the flue gas ducts, especially as plants age.

As a CO_2 capture method, oxy-combustion has been studied theoretically and in experimental laboratory and pilot plant facilities, but not yet at a commercial scale. Thus, a variety of designs have been proposed for commercial systems.[7] Although in principle oxyfuel systems can capture all of the CO_2 produced, the need for additional gas treatment systems decreases the capture efficiency to about 90% in most current designs.

In principle, oxy-combustion also can be applied to simple cycle and combined cycle power plants fueled by natural gas or distillate oil. These conceptual designs are discussed more fully in "Chapter 7: Status of Oxy-Combustion Capture." As a practical matter, however, they would require significant and costly modifications to the design of current gas turbines and other plant equipment, with relatively limited market potential for greenhouse gas abatement. Thus, the current focus of oxy-combustion development is on coal-fired power plant applications.

Capture System Energy Penalty

The energy requirements of current CO_2 capture systems are roughly 10 to 100 times greater than those of other environmental control systems employed at a modern electric power plant. This energy "penalty" lowers the overall (net) plant efficiency and significantly increases the net cost of CO_2 capture. **Table 2** shows that of the three CO_2 capture approaches discussed earlier, post-combustion capture on PC plants is the most energy-intensive, requiring nearly twice the energy per net unit of electricity output as pre-combustion capture on an IGCC plant.

Table 2. Representative Values of Current Power Plant Efficiencies and CCS Energy Penalties

Power plant type, and capture system type	Net plant efficiency (%) without CCS	Net plant efficiency (%) with CCS	Energy penalty: Added fuel input (%) per net kWh output
Existing subcritical PC, post-combustion capture	33	23	40%
New supercritical PC, post-combustion capture	40	31	30%
New supercritical PC, oxy-combustion capture	40	32	25%
New IGCC (bituminous coal), pre-combustion capture	40	34	19%
New natural gas combined cycle, post-combustion capture	50	43	16%

Sources: Metz, *Special Report*; Massachusetts Institute of Technology (MIT), *The Future of Coal* (Cambridge, MA: MIT, 2007); Carnegie Mellon University, *Integrated Environmental Control Model (IECM)*, December 2009.

a.   All efficiency values are based on the higher heating value (HHV) of fuel.

Notes: For each plant type, there is a range of efficiencies around the representative values shown here.

Lower plant efficiency means that more fuel is needed to generate electricity relative to a similar plant without CO_2 capture. For coal combustion plants, this means that proportionally more solid waste is produced and more chemicals, such as ammonia and limestone, are needed (per unit of

---

[7] Metz, *Special Report*.

electrical output) to control NO_x and SO_2 emissions. Plant water use also increases significantly because of the additional cooling water needed for current amine capture systems. Because of the efficiency loss, a capture system that removes 90% of the CO_2 from the plant flue gas winds up reducing the net (avoided) emissions per kilowatt-hour (kWh) by a smaller amount, typically 85% to 88%.

In general, the higher the power plant efficiency, the smaller is the energy penalty and its associated impacts. For this reason, replacing or repowering an old, inefficient plant with a new, more efficient unit with CO_2 capture can still yield a net efficiency gain that decreases all plant emissions and resource consumption. Thus, the net impact of the CO_2 capture energy penalty must be assessed in the context of a particular situation or strategy for reducing CO_2 emissions. Innovations that raise the efficiency of power generation also can reduce the impacts and cost of carbon capture. **Table 3** shows that the overall energy requirements for PC and IGCC plants is divided between electricity needed to operate fans, pumps, and CO_2 compressors, plus thermal energy requirements (or losses) for solvent regeneration (PC plants) and the water-gas shift reaction (IGCC plants). Thermal energy requirements are clearly the largest source of net power losses and the priority area for research to reduce those losses. For oxy-combustion systems, the electrical energy required for oxygen production is the biggest contributor to the energy penalty.

Table 3. Breakdown of the Energy Penalty for CO_2 Capture at Supercritical PC and IGCC Power Plants

Energy Type and Function	Approximate % of Total Energy Penalty
Thermal energy for amine solvent regeneration (post-combustion) or loss in water-gas shift reaction (pre-combustion); or, electricity for oxygen production (oxy-combustion)	~60%
Electricity for CO_2 compression	~30%
Electricity for pumps, fans, etc.	~10%

Sources: MIT, "Future of Coal"; Carnegie Mellon, "IECM."

Current Cost of CO_2 Capture

To gauge the potential benefits of advances in carbon capture technology, it is useful to first benchmark the cost of current systems. This section reviews recent cost estimates for power plants and other industrial processes.

Costs for New Power Plants

Figure 9 displays the cost of generating electricity from new power plants with and without CCS, as reported in recent studies based on current commercial post-combustion and pre-combustion capture processes. All plants capture and sequester 90% of the CO_2 in deep geologic formations.

The total cost of electricity generation (COE), in dollars per megawatt-hour ($/MWh), is shown as a function of the CO_2 emission rate (tonnes CO_2/MWh) for power plants burning bituminous coal or natural gas. The COE includes the costs of CO_2 transport and storage, but most of the cost (80% to 90%) is for capture (including compression).

Figure 9. Cost of Electricity Generation (2007 US$/MWh) as a Function of the CO₂ Emission Rate (tonnes CO₂/MWh) for New Power Plants Burning Bituminous Coal or Natural Gas

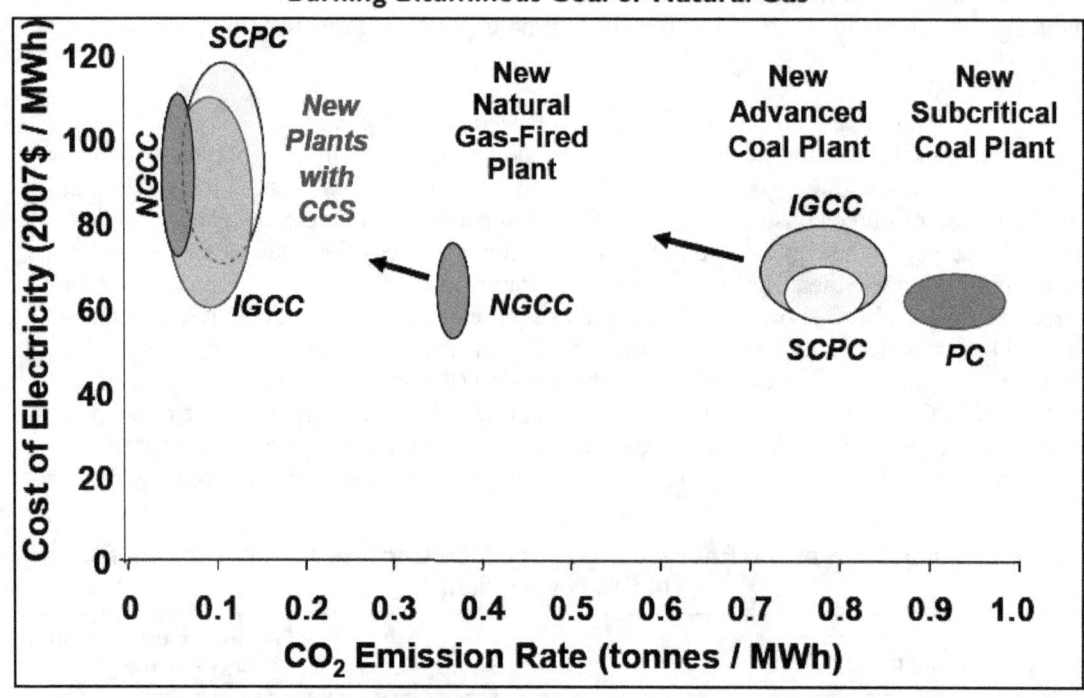

Source: Adapted from E. S. Rubin, "CO₂ Capture and Transport," *Elements*, vol. 4, no. 5 (2008), pp. 311-317.

Notes: PC = subcritical pulverized coal units; SCPC = supercritical pulverized coal; IGCC = integrated gasification combined cycle; NGCC = natural gas combined cycle). Ranges reflect differences in technical and economic parameters affecting plant cost, based on data from DOE, "Cost and Performance"; N. Holt, "CO₂ Capture & Storage—EPRI CoalFleet Program," PacificCorp Energy IGCC/Climate Change Working Group, January 25, 2007, Electric Power Research Institute, Palo Alto, CA; MIT, 2007; E. S. Rubin, C. Chen, and A. B. Rao, "Cost and Performance of Fossil Fuel Power Plants with CO₂ Capture and Storage," *Energy Policy*, vol. 35, no. 9 (2007), pp. 4444-4454; and Metz, 'Special Report.'

The dominant factors responsible for the broad range of costs for each plant type in **Figure 9** are assumptions about the design, operation, and financing of the power plant to which the capture technology is applied. For example, higher plant efficiency, larger plant size, higher fuel quality, lower fuel cost, higher annual hours of operation, longer operating life, and lower cost of capital all reduce both the cost of electricity and the unit cost of CO_2 capture. Assumptions about the CO_2 capture system design and operation also contribute to variations in the overall cost. Assumptions vary across the set of studies cited. Since no single set of assumptions applies to all situations or all parts of the world, there is no single estimate for the cost of CO_2 capture. Indeed, the cost ranges would be even broader if other factors such as a larger range of boiler efficiencies or coal types were considered.

On a relative basis, CCS is estimated to increase the cost of generating electricity by approximately 60% to 80% at new coal combustion plants and by about 30% to 50% at new coal gasification plants. On an absolute basis, the increased cost translates to roughly $40-$70/MWh for supercritical (SCPC) coal plants and $30-$50/MWh for IGCC plants using bituminous coal. As noted earlier, the CO_2 capture step (which includes CO_2 compression) accounts for 80% to 90% of this cost.

Figure 9 also can be used to calculate the cost per tonne of CO_2 avoided for a plant with capture relative to one without. This cost is equivalent to the "carbon price" or CO_2 emissions tax above which the CCS plant is more economical than the plant without capture. For new supercritical coal plants this is currently about \$60-\$80/tonne CO_2. For IGCC plants with and without CCS, the avoidance cost is smaller, about \$30-\$50/tonne CO_2. Since the cost of CO_2 avoided depends on the choice of "reference plant" with no CCS, it is also useful to compare an IGCC plant with capture to a SCPC reference plant without capture. In this case, the cost of CO_2 avoided is roughly \$40-\$60/tonne CO_2. In all cases, costs are lower if the CO_2 can be sold for enhanced oil recovery (EOR) with subsequent geological storage. For plants using low-rank coals (i.e., subbituminous coal or lignite), the avoidance cost may be slightly higher.[8]

Retrofit Costs for Existing Power Plants

For existing power plants, the feasibility and cost of retrofitting a CO_2 capture system depend heavily on site-specific factors such as the plant size, age, efficiency, type and design of existing air pollution control systems, and availability of space to accommodate a capture unit.[9] In general, the added cost of electricity generation is higher than for a new supercritical plant. A major contributing factor is the lower thermal efficiency typical of existing (subcritical) power plants, which results in a larger energy penalty and higher capital cost per unit of capacity. Other factors include the added capital costs due to physical constraints and site access difficulties during construction of a retrofit project, plus the likely need for upgrades or installation of additional equipment, such as more efficient SO_2 scrubbers. The cost per ton of CO_2 avoided also increases as a result of these higher costs.

Studies also indicate that for many existing plants the most cost-effective strategy for plants that have suitable access to geological storage areas is to combine CO_2 capture with a major plant upgrade, commonly called repowering. Here, an existing subcritical unit is replaced either by a high-efficiency (supercritical) boiler and steam turbine system, or by a gasification combined cycle system.[10] In such cases, the cost of CO_2 capture approaches that of a new plant, with some potential savings from the use of existing plant components and infrastructure, as well as from fewer operating permit requirements relative to a new greenfield site.

Costs for Other Industrial Processes

There have been a limited number of studies of CO_2 capture costs for industrial processes other than power plants. **Table 4** summarizes the reported cost ranges.[11] In general, the incremental cost of capture is lowest for processes where CO_2 is already separated as part of the normal process operations, such as in the production of hydrogen or the purification of natural gas. In these cases,

---

[8] U.S. Department of Energy, *Assessment of Power Plants That Meet Proposed Greenhouse Gas Emission Performance Standards* DOE/NETL-401/110509, National Energy Technology Laboratory, Pittsburgh, PA, November 5, 2009; E. S. Rubin, C. Chen, and A. B. Rao, "Cost and Performance of Fossil Fuel Power Plants with CO_2 Capture and Storage," *Energy Policy*, vol. 35, no. 9 (2007), pp. 4444-4454.

[9] Rao and Rubin, "Technical, Economic."

[10] C. Chen, A. B. Rao, and E. S. Rubin, "Comparative Assessment of CO_2 Capture Options for Existing Coal-Fired Power Plants," Proc. *Second National Conference on Carbon Sequestration, May 5-8, 2003,* Alexandria, VA.; D. Simbeck, "The carbon capture technology landscape," Proc. *Energy Frontiers International Emerging Energy Technology Forum,* SFA Pacific, Inc., February, 2008, Mountain View, CA.

[11] Metz, *Special Report.*

---

the added cost is simply for CO_2 compression. For other industrial processes, capture costs are highly variable and depend strongly on site-specific factors, both technical and economic.

Table 4. Range of CO_2 Capture Costs for Several Types of Industrial Processes
(2007$/tonne CO_2)

Industrial Process	Capture Cost Range
Fossil fuel power plants	$20-$95/t CO_2 net captured
Hydrogen and ammonia production, or a natural gas processing plant	$5-$70/t CO_2 net captured
All other industrial processes	$30-$145/t CO_2 net captured

Source: Based on Metz, *Special Report* data, adjusted to 2007 cost basis.

Important Caveat Concerning Costs

Construction costs for power plants and industrial equipment escalated dramatically from about 2004 to 2008, as did fuel prices, especially natural gas. Most prices then stabilized or receded during the subsequent economic recession. Uncertainty about future cost trends, together with the absence of full-scale projects, further clouds the "true" cost of facilities with or without CCS. For power plants, the relative costs of PC and IGCC plants also can change with coal type, operating hours, cost of capital, and many other factors.[12] Experience with IGCC power plants is still quite limited, and neither PC nor IGCC plants with CCS have yet been built and operated at full scale. Thus, neither the absolute nor the relative costs of these systems can be stated with a high degree of confidence at this time.

---

[12] Rubin et al., "Cost and Performance."

Chapter 4: Stages of Technology Development

Introduction

The stages of technological development or maturity of carbon capture systems span a broad spectrum. At one end of the spectrum are the current commercial systems described in the previous chapter. At the opposite end are new concepts or processes that exist only on paper, or perhaps as a small-scale device or experiment in a research laboratory. New or "advanced" technologies commonly seek (and often boast of) higher effectiveness and/or lower cost than current commercial systems—attributes that are highly desired in the marketplace. At the same time, claims about the cost or performance of processes in the early stages of development are inherently uncertain and subject to change as the technology advances toward commercialization.

This chapter discusses a number of ways to characterize the level of technological development of CO_2 capture systems. The aim is to provide a clear understanding of the steps that are needed to bring a promising new technology to commercial reality. To begin, however, this section briefly describes the general process of technological change in order to provide context for a closer examination of innovations in carbon capture technologies.

The Process of Technological Change

Innovations in carbon capture technology and the commercial adoption of such systems are examples of the general process of technological change. While a variety of terms are used to describe that process, four commonly defined stages are:

- **Invention**—discovery; creation of knowledge; new prototypes

- **Innovation**—creation of a new commercial product or process

- **Adoption**—deployment and initial use of the new technology

- **Diffusion**—increasing adoption and use of the technology

The first stage is driven by R&D, including both basic and applied research. The second stage—innovation—is a term often used colloquially to describe the overall process of technological change. As used here, however, it refers only to the creation of a product or process that is commercially offered; it does not mean the product will be adopted or become widely used. That happens only if the product succeeds in the final two stages—adoption and diffusion, which reflect the commercial success of a technology innovation.

Studies also show that rather than being a simple linear process, the four stages of technological change are highly interactive, as depicted in **Figure 10**. Thus, innovation is stimulated not only by support for R&D, but also by the experience of early adopters, plus added knowledge gained as a technology diffuses more widely into the marketplace. The reductions in product cost that are often observed as a technology matures—commonly characterized as a "learning curve"—reflect the combined impacts of sustained R&D plus the benefits derived from "learning by doing" (economies in the manufacture of a product) and "learning by using" (economies in the operating costs of a product).

Figure 10. Stages of Technological Change and Their Interactions

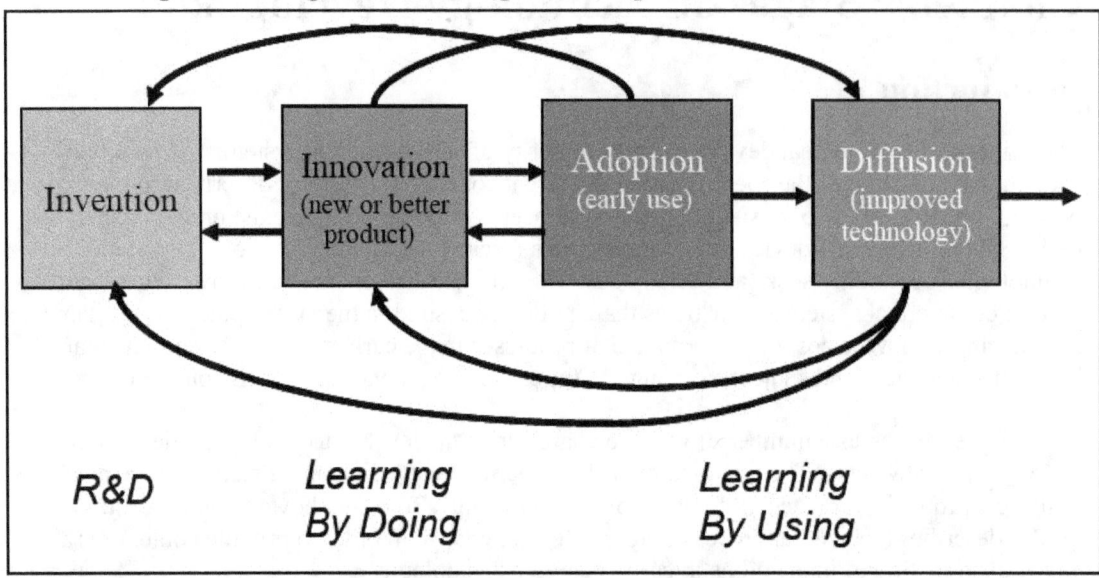

Source: E. S. Rubin, "The Government Role in Technology Innovation: Lessons for the Climate Change Policy Agenda," Institute of Transportation Studies, 10th Biennial Conference on Transportation Energy and Environmental Policy, University of California, Davis, CA (August 2005).

This report deals only with the first two stages of **Figure 10** in the context of carbon capture systems at different levels of development or maturity. The goal is to characterize the current status of capture technologies and the outlook for future commercial systems. Later, "Chapter 9: Lessons from Past Experience" discusses the influence of the last two stages (adoption and diffusion) on the pace of innovation and the prospects for lower-cost capture technologies.

Technology Readiness Levels (TRLs)

One method of describing the maturity of a technology or system is the scale of technology readiness levels (TRLs) depicted in **Figure 11**. First developed for the National Aeronautics and Space Administration (NASA), TRLs were subsequently adopted by the U.S. Department of Defense, as well as by other organizations involved in developing and deploying complex technologies or systems, both in the United States and abroad. Recently, researchers at the Electric Power Research Institute (EPRI) also adopted TRLs to describe the status of new post-combustion carbon capture technologies,[13] discussed later in "Chapter 5: Status of Post-Combustion Capture."

---

[13] Electric Power Research Institute, *Program on Technology Innovation: Post-Combustion CO₂ Capture Technology Development*, Report No. 1016995, Prepared by A. S. Bhown and B. Freeman, Palo Alto, CA, December 2008.

Figure 11. Descriptions of Technology Readiness Levels (TRLs)

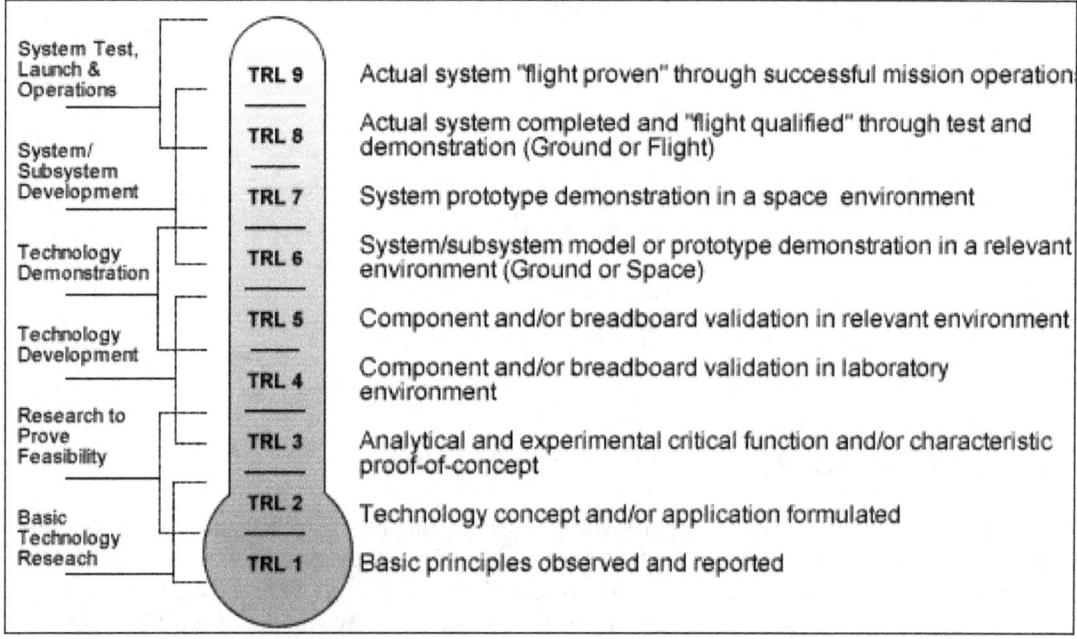

Source: National Aeronautics and Space Administration, "Definition of Technology Readiness Levels," at http://esto.nasa.gov/files/TRL_definitions.pdf.

The TRL scale has nine levels. At TRL 1 a technology consists only of basic principles, while at TRL 9 it has evolved into a system that has been used successfully in its actual operating environment. TRLs are used to assess the maturity of a technology and the risks of placing it into service for a given mission. Studies by the U.S. Government Accountability Office (GAO) found that commercial firms typically do not introduce new technology into a commercial product until it is at the equivalent of TRL 8 or TRL 9, where the technology has been fully integrated and validated in its working environment. The GAO also found that a number of government projects it examined tended to be further behind schedule and over budget where unproven technologies were employed, compared to projects designed with more mature technologies.[14]

DOE's Office of Management also recently published a *Technology Readiness Assessment Guide* to provide general guidance as to how critical technologies should be developed before and during their integration into engineered systems.[15] The modified definitions of TRLs employ four scales of development called lab scale, bench scale, engineering scale and full scale (**Figure 12**). A technology is considered to be lab scale at TRLs 2 and 3 and bench scale at TRL 4. The latter is typically a complete system, whereas lab scale involves proof-of-concept for a subsystem or component. A technology at the engineering scale corresponds to TRLs 5 and 6. At TRL 7 and beyond the system is full scale. Variants of these four categories are used in this report to describe the development stages of carbon capture technologies, as explained below.

---

[14] U.S. General Accounting Office, *Better Management of Technology Development Can Improve Weapon System Outcomes*, GAO/NSIAD-99-162, Washington, DC (July 1999); Government Accountability Office, *Major Construction Projects Need a Consistent Approach for Assessing Technology Readiness to Help Avoid Cost Increases and Delays*, Washington, DC (July 2007).

[15] U.S. Department of Energy, *Technology Readiness Assessment Guide,* at http://www.directives.doe.gov/directives/current-directives/413.3-EGuide-04/view?searchterm=None.

Figure 12. A Department of Energy View of Technology Development Stages and Their Corresponding TRLs

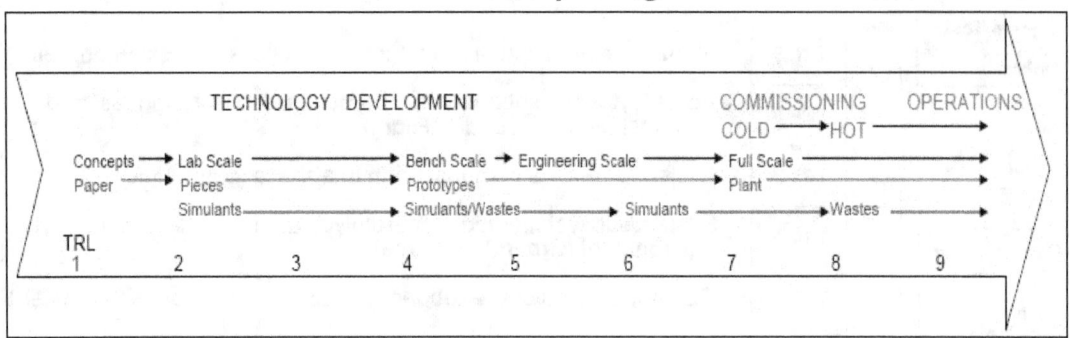

Source: DOE, "Technology Readiness."

Technology Maturity Levels Used in this Study

While the nine-level TRL scale is a useful way to describe and compare the status of technologies being considered for deployment in a particular mission or complex system, for purposes of this study, a simpler set of five categories is used to describe the maturity of carbon capture technologies. The five stages reflect not only different levels of maturity but also differences in the physical size and complexity of a technology at different points in its development. Significant increases in the level of financial commitments also are needed to advance along this five-stage journey, which not all processes survive. This representation of "what's in the pipeline" is possibly the most effective way to convey to Congress and others the prospects, time requirements, and level of financial resources needed to bring improved CO_2 capture systems to the marketplace.

Commercial Process

A commercial carbon capture technology or process is one that is available for routine use in a particular application such as a power plant or industrial process. The capture technology is offered for sale by one or more reliable vendors with standard commercial guarantees. As defined here, a commercial technology corresponds to TRL 9, the highest level on the TRL scale. This is the maturity level that electric utility companies normally will require before installing a carbon capture system at a U.S. power plant.

Full-Scale Demonstration Plant

The full-scale demonstration stage corresponds to levels 7 and 8 on the TRL scale. It represents the stage at which a CO_2 capture technology is integrated into a full-size system in order to demonstrate its viability and commercial readiness in a particular application. For power plants, such applications might include pulverized coal combustion systems employing oxy-combustion or post-combustion CO_2 capture, as well as IGCC plants employing pre-combustion capture. While there is flexibility in the definition of "full-scale," in general a full-scale demonstration would correspond to a gross power plant size of approximately 250 MW, with a corresponding CO_2 capture rate of roughly 1-2 million tonnes per year for a coal-fired plant. For reference, the median size of U.S. coal-burning power plants today is approximately 650 MW (gross or nameplate capacity). For gas-fired power plants or other industrial applications a full-scale demonstration may have smaller annual quantities of CO_2 captured because of smaller plant sizes and/or lower fuel carbon content.

Pilot Plant Scale

The pilot plant stage is where a process or technology is tested in a realistic environment, but at a scale that is typically one to two orders of magnitude smaller than the full-scale demonstration. For carbon capture processes, a pilot plant might be built as a stand-alone facility, or as a unit capturing CO_2 from the slipstream of an adjoining full-size power plant. Pilot plants represent an initial demonstration stage corresponding to levels 6 and 7 on the TRL scale. At this stage data are gathered to refine and further develop a process, or to design a full-size (or intermediate size) demonstration plant.

Laboratory or Bench Scale

The laboratory and bench scales represent the early stage of process development in which an apparatus or process is first successfully constructed and operated in a controlled environment, often using materials and test gases to simulate a commercial process or stream (such as a flue gas stream). A bench-scale apparatus is typically built as a complete representation of a process or system, whereas laboratory-scale experiments typically seek to validate or obtain data for specific components of a system. Laboratory- and bench-scale processes correspond to levels 3, 4 and 5 on the TRL scale.

Conceptual Design

The conceptual design stage of a CO_2 capture process is one for which the basic science has been developed, but no physical prototypes yet exist. Conceptual designs are often developed and tested with computer models before any laboratory work is done. This allows for confirmation that the design principles are sound, plus some degree of process optimization before progressing to the more expensive laboratory or bench-scale stage. The conceptual design stage corresponds to levels 1 and 2 on the TRL scale.

Current Status of CO_2 Capture Technologies

Consistent with the objectives of this study, the next three chapters characterize the current status of carbon capture technologies with respect to the five stages of development outlined above. Each chapter addresses one of the three main avenues for CO_2 capture—post-combustion, pre-combustion, and oxy-combustion systems. The subsequent chapter then discusses the cost reductions expected from advanced capture systems and the projected timetables for their commercialization.

Chapter 5: Status of Post-Combustion Capture

Introduction

This chapter summarizes the status of post-combustion CO_2 capture technologies at various stages of development. The most advanced systems today employ amine-based solvents, while processes at the earliest stages of development employ a variety of novel solvents, solid sorbents, and membranes for CO_2 capture or separation. The chapter begins with a summary of current commercial processes and then describes technologies at each of the four other stages of development defined in "Chapter 4: Stages of Technology Development."

In recent years, carbon capture R&D programs have expanded rapidly throughout the world; thus, any summary of "current" activities and projects is soon out of date. For this reason, this report does not attempt to cover capture-related R&D activities comprehensively. Rather, it attempts to synthesize key findings from our own investigations and from the work of others who also track and report on the status of CO_2 capture technology developments. It draws also upon a set of publicly available databases and CCS project status reports maintained by organizations including DOE's National Energy Technology Laboratory (DOE/NETL), the International Energy Agency's Greenhouse Gas Control Programme (IEAGHG), the Massachusetts Institute of Technology (MIT) Carbon Sequestration Program, and the recently formed Global Carbon Capture and Storage Institute (GCCSI).[16] In some cases, the information from these public databases has been supplemented by data from websites of companies involved in capture technology development.

In each of the sections below, the objective is to summarize not only the current status of post-combustion capture technology developments (as of March 2010), but also the potential advantages of each new technology, as well as the key technical barriers and challenges that must be overcome to advance the method. Brief descriptions of new processes or capture methods not previously discussed in "Chapter 3: Overview of CO_2 Capture Technologies" also are provided.

Commercial Processes

As noted in "Chapter 3: Overview of CO_2 Capture Technologies," post-combustion CO_2 capture systems have been in use commercially for many decades, mainly in industrial processes for purifying gas streams other than combustion products. The use of amines to capture CO_2 was first patented 80 years ago and since then has been used to meet CO_2 product specifications in industries ranging from natural gas production to the food and beverage industry.[17] A number of vendors currently offer commercial amine-based processes, including the Fluor Daniel Econamine FG Plus process, the Mitsubishi Heavy Industries KM-CDR process, the Lummus Kerr-McGee process, the Aker Clean Carbon Just Catch process, the Cansolv CO_2 capture system, and the HTC Purenergy Process.[18]

---

[16]U.S. Department of Energy, "NETL Carbon Capture and Storage Database," http://www.netl.doe.gov/technologies/ carbon_seq/database/index html; International Energy Agency Greenhouse Gas R&D Programme (IEAGHG), "CO_2 Capture and Storage." http://www.co2captureandstorage.info/co2db.php; MIT Energy Initiative, "Carbon Capture and Sequestration Technologies at MIT," http://sequestration mit.edu; Global Carbon Capture and Storage Institute (GCCSI), *Strategic Analysis of the Global Status of Carbon Capture and Storage*, WorleyParsons, 2009, http://www.globalccsinstitute.com/downloads/Status-of-CCS-WorleyParsons-Report-Synthesis.pdf.

[17] G. Rochelle, "Amine Scrubbing for CO_2 Capture," *Science*, vol. 325 (2009), pp. 1652-1654.

[18] Clean Air Task Force & Consortium for Science, Policy and Outcomes, "Innovation Policy for Climate Change," Proc. *National Commission on Energy Policy*, Washington, DC.

The hundreds of commercial aqueous amine systems currently in operation typically vent the captured CO_2 to the atmosphere. Of the projects listed in **Table 5**, three are at natural gas treatment plants (two in Norway, one in Algeria) in which the captured CO_2 is sequestered in deep geological formations to prevent its release to the atmosphere. One of these projects, the Statoil natural gas production facility at Sleipner in the North Sea, has been operating since 1996. This is the longest-running commercial CCS project. **Figure 13** shows a photograph of the amine-based CO_2 capture unit installed more recently at a natural gas treatment plant in Algeria. That unit is part of an integrated CCS system that includes CO_2 capture, pipeline transport, and sequestration in a nearby geological formation.

Table 5. Commercial Post-Combustion Capture Processes at Power Plants and Selected Industrial Facilities

Project Name and Location	Plant and Fuel Type	Year of Startup	Approx. Capture Plant Capacity	Capture System Type (Vendor)	CO_2 Captured (10^6 tonnes/yr)
United States					
IMC Global Inc. Soda Ash Plant (Trona, CA)	Coal and petroleum coke-fired boilers	1978	43 MW	Amine (Lummus)	0.29
AES Shady Point Power Plant (Panama City, OK)	Coal-fired power plant	1991	9 MW	Amine (Lummus)	0.06
Bellingham Cogeneration Facility (Bellingham, MA)	Natural gas-fired power plant	1991	17 MW	Amine (Fluor)	0.11
Warrior Run Power Plant (Cumberland, MD)	Coal-fired power plant	2000	8 MW	Amine (Lummus)	0.05
Outside the United States					
Soda Ash Botswana Sua Pan Plant (Botswana)	Coal-fired power plant	1991	17 MW	Amine (Lummus)	0.11
Sumitomo Chemicals Plant (Japan)	Gas & coal boilers	1994	8 MW	Amine (Fluor)	0.05
Statoil Sleipner West Gas Field (North Sea, Norway)	Natural gas separation	1996	N/A	Amine (Aker)	1.0
Petronas Gas Processing Plant (Kuala Lumpur, Malaysia)	Natural gas-fired power plant	1999	10 MW	Amine (MHI)	0.07
BP Gas Processing Plant (In Salah, Algeria)	Natural gas separation	2004	N/A	Amine (Multiple)	1.0
Mitsubishi Chemical Kurosaki Plant (Kurosaki, Japan)	Natural gas-fired power plant	2005	18 MW	Amine (MHI)	0.12
Snøhvit Field LNG and CO_2 Storage Project (North Sea, Norway)	Natural gas separation	2008	N/A	Amine (Aker)	0.7
Huaneng Co-Generation Power Plant (Beijing, China)	Coal-fired power plant	2008	0.5 MW	Amine (Huaneng)	0.003

Sources: DOE, "NETL Carbon"; IEAGHG, "CO_2 Capture"; MIT, "Carbon Capture"; GCCSI, "Strategic Analysis."

Figure 13. An Amine-Based CO$_2$ Capture System Used to Purify Natural Gas at BP's In Salah Plant in Algeria

Source: Photo courtesy of IEA Greenhouse Gas Programme.

As shown in **Table 5**, CO$_2$ is also captured at several coal-fired and gas-fired power plants where a portion of the flue gas stream is fitted with a CO$_2$ capture system. **Figure 14** shows the amine systems installed at two U.S. power plants, one burning coal, the other natural gas. The CO$_2$ captured at these plants is sold to nearby food processing facilities, which use it to make dry ice or carbonated beverages. The oldest and largest commercial CO$_2$ capture system operating on flue gases is the IMC Global soda ash plant in California. Here, the mineral trona is mined locally and combined with CO$_2$ to produce sodium carbonate (soda ash), a widely used industrial chemical.[19] All these products soon release the CO$_2$ to the atmosphere (e.g., through carbonated beverages).

To date, only ABB Lummus (now CB&I Lummus) has commercial flue gas CO$_2$ capture units operating at coal-fired power plants, while both Fluor Daniel and MHI have commercial installations at gas-fired plants (see **Table 5**). Both Fluor and MHI now also offer commercial guarantees for post-combustion capture at coal-fired power plants.

These vendors (and others) use amine-based solvents for CO$_2$ capture. In most cases the exact composition of the solvent is proprietary. The currently operating Lummus systems employ a solution of 20% MEA in water, while the Fluor systems use a solvent with a 30% amine concentration.[20] Higher amine concentrations are beneficial in reducing the large energy penalty of CO$_2$ capture, since there is less water in the solution that needs to be pumped and heated in the regeneration process. Capital cost is also less, since higher amine concentrations lead to smaller equipment sizes. On the other hand, amines such as MEA are highly corrosive, so higher amine concentrations require chemical additives or more costly construction materials to prevent corrosion. Tradeoffs among these factors underlie some of the differences in capture system designs offered by different vendors. The systems and solvents currently offered commercially by Fluor (Econamine FG+) and MHI (KS-1) boast of reductions of roughly 25% in capture energy requirements relative to older system designs using MEA, which lowers the overall cost.

---

[19] IEAGHG, "CO$_2$ Capture."

[20] P. H. M. Feron, "Progress in SP2 CO$_2$ Post-combustion Capture," Presentation at ENCAP/CASTOR Seminar, March 2006. J. N. Jensen and J. N. Knudsen, "Experience with the CASTOR/CESAR Pilot Plant," presentation at the Workshop on Operating Flexibility of Power Plants with CCS, November 2009.

Figure 14. Amine-Based Post-Combustion CO$_2$ Capture Systems Treating a Portion of the Flue Gas from a Coal-Fired Power Plant in Oklahoma, USA (left), and a Natural Gas Combined Cycle (NGCC) Plant in Massachusetts, USA (right)

Source: Photos courtesy of ABB Lummus, Fluor Daniels, and Chevron.

Full-Scale Demonstration Plants

Although several CO$_2$ capture systems have operated commercially for nearly two decades on a portion of power plant flue gases, no capture units have yet been applied to the full flue gas stream of a modern coal-fired or gas-fired power plant. Thus, one or more demonstrations of post-combustion CO$_2$ capture at full scale are widely regarded as crucial for gaining the acceptance of this technology by electric utility companies, as well as by the institutions that finance and regulate power plant construction and operation. Several years ago, for example, the European Union called for 12 such demonstrations in Europe, while in the United States there have been calls for at least 6 to 10 full-scale projects.[21]

To date, however, no such demonstrations have yet occurred, nor (as best we can tell) has full financing yet been guaranteed for any of the full-scale demonstration projects that have been announced. One reason is the high cost of each project, estimated at roughly $1 billion for CO$_2$ capture at a 400 MW unit operating for five years.[22] Several previously announced demonstrations of full-scale power plant capture and storage systems have been canceled or delayed due to sharp escalations in construction costs prior to 2008. Even more recently, a 160 MW demonstration project in the United States was canceled not long after being announced.[23]

---

[21] European Technology Platform for Zero Emission Fossil Fuel Power Plants, "EU Demonstration Programme for CO$_2$ Capture and Storage (CCS)," November 2008, http://www.zeroemissionsplatform.eu/index; MIT, 'Future of Coal.' V.A. Kuuskraa, "A Program to Accelerate the Deployment of CO$_2$ Capture and Storage (CCS): Rationale, Objectives and Costs," Paper prepared for the Coal Initiative Reports' Series of the Pew Center on Global Climate Change, Arlington, VA, October 2007.

[22] Kuuskraa, "Accelerate Deployment."

[23] Sourcewatch, "Southern Company abandons carbon capture and storage project," 2010, http://www.sourcewatch.org/index.php?title=Southern_Company#cite_note-11.

Nevertheless, it appears reasonable to assume that at least some of the large-scale projects currently planned for post-combustion CO_2 capture in the United States and other countries will materialize over the next several years, with costs shared between the public and private sectors.

Table 6 lists the features and locations of major announced demonstration projects at power plants in the United States and other countries. Most of these CO_2 capture systems would be installed at existing coal-fired plants, with the captured CO_2 transported via pipeline to a geological storage site (often in conjunction with enhanced oil recovery to reduce project costs).

Table 6. Planned Demonstration Projects at Power Plants with Full-Scale Post-Combustion Capture

Project Name and Location	Plant and Fuel Type	Year of Startup	Approx. Capture Plant Capacity	Capture System Type (Vendor)	Annual CO₂ Captured (10⁶ tonnes)	Current Status (March 2010)
United States						
Basin Electric Antelope Valley Station (Beulah, ND)	Coal-fired power plant	2012	120 MW	Amine (HTC)	1.0	Site Selection
Tenaska Trailblazer Energy Center (Sweetwater, TX)	Coal-fired power plant	2014	600 MW	Amine (Fluor)	4.3	Permitting
American Electric Power Mountaineer Plant (New Haven, WV)	Coal-fired power plant	2015	235 MW	Chilled Ammonia (Alstom)	1.5	Scoping
Outside the United States						
SaskPower Boundary Dam Polygon (Estevan, Canada)	Coal-fired power plant	2014	115 MW	Amine (Cansolv)	1.0	Plant Design
E.ON Kingsnorth Ruhrgas UK Post-Combustion Project (Kent, United Kingdom)	Coal-fired power plant	2014	300 MWᵃ	Amine (Fluor & MHI)	1.9	Plant Design
TransAlta Project Pioneer Keephills 3 Power Plant (Wabamun, Canada)	Coal-fired power plant	2015	200 MW	Chilled Ammonia (Alstom)	1.0	Plant Design
Vattenfall Janschwalde (Janschwalde, Germany)	Coal-fired power plant	2015	125 MW	Amine (TBD)	—	Permitting
Porto Tolle (Rovigo, Italy)	Coal-fired power plant	2015	200 MWᵃ	Amine (TBD)	1.0	Scoping

Source: DOE, "NETL Carbon'"; IEAGHG, "CO₂ Capture"; MIT, "Carbon Capture"; GCCSI, "Strategic Analysis."

a.   Estimated from other reported data.

Note that while most of the projects in **Table 6** plan to employ amine-based capture systems, a few propose to use an ammonia-based process. Two such capture processes are currently at the pilot plant stage and are described in more detail in the next section of this report. Plans for scale-up to a demonstration project are predicated on successful operation of the smaller-scale pilot plants.

Note too that most of the planned demonstration projects have expected startup dates of 2014 or later. This means that such projects are currently in the early stages of detailed design and that final commitments of full funding for construction have not yet been made. Similarly, it is still too early to know the details of capture system designs and the extent to which they might be expected to achieve further improvements in CO_2 capture efficiency and/or reductions in cost relative to current commercial systems.

In addition to the power plant projects in **Table 6**, DOE plans to support at least four demonstration projects of CO_2 capture at industrial facilities. Eleven candidates were selected for further study in late 2009, with down-selections expected in mid-2010.

Pilot Plant Projects

Table 7 lists pilot-scale post-combustion CO_2 capture projects that are currently operating or are in the design or construction stage. Most of these projects are testing and developing new or improved amine-based solvents. Several others are testing and developing ammonia-based capture processes.

Table 7. Pilot Plant Processes and Projects for Post-Combustion CO_2 Capture

Project Name and Location	Plant and Fuel Type	Year of Startup	Approx. Capture Plant Capacity	Capture System Type (Vendor)	Annual CO_2 Captured (10^6 tonnes)
United States					
First Energy R.E. Burger Plant (Shadyside, OH)	Coal-fired power plant	2008	1 MW	Ammonia (Powerspan)	0.007
American Electric Power Mountaineer Plant (New Haven, WV)	Coal-fired power plant	2009	20 MW	Chilled Ammonia (Alstom)	0.1
Dow Chemicals South Charleston Plant (Charleston, WV)	Coal-fired power plant	2009	0.5 MW[a]	Amines (Dow/ Alstom)	0.002
NRG Energy WA Parish Plant (Houston, TX)	Coal-fired power plant	2012	60 MW	Amine (Fluor)	0.5
Outside the United States					
Nanko Natural Gas Pilot Plant (Osaka, Japan)	Gas-fired power plant	1991	0.1 MW	Amine (MHI)	0.001
Matsushima Coal Plant (Nagasaki, Japan)	Coal-fired power plant	2006	0.8 MW[a]	Amine (MHI)	0.004

Project Name and Location	Plant and Fuel Type	Year of Startup	Approx. Capture Plant Capacity	Capture System Type (Vendor)	Annual CO_2 Captured (10^6 tonnes)
Munmorah Pilot Plant (Lake Munmorah, Australia)	Coal-fired power plant	2008	1 MW[a]	Ammonia (Delta & CSIRO)	0.005
Tarong Power Station (Nanango, Australia)	Coal-fired power plant	2008	0.5 MW[a]	Amine (Tarong & CSIRO)	0.0015
Hazelwood Carbon Capture (Morewell, Australia)	Coal-fired power plant	2008	2 MW	Amine (Process Group)	0.01
CASTOR CO_2 from Capture to Storage (Esbjerg, Denmark)	Coal-fired power plant	2008	3 MW	Amine (Multiple)	0.008
Eni and Enel Federico II Brindisi Power Plant (Brindisi, Italy)	Coal-fired power plant	2009	1.5 MW	Amine (Enel)	0.008
CATO-2 CO_2 Catcher (Rotterdam, Netherlands)	Coal-fired power plant	2008	0.4 MW	Amine (Multiple)	0.002
Statoil Mongstad Cogeneration Pilot (Mongstad, Norway)	Natural gas-fired power plant	2010	15 MW[a]	Chilled NH_3 (Alstom)	0.08
			7 MW[a]	Amine (Various)	0.02

Sources: DOE, "NETL Carbon"; IEAGHG, "CO_2 Capture"; MIT, "Carbon Capture"; GCCSI, "Strategic Analysis."

a.   Estimated from other reported data.

Amine-Based Capture Processes

The class of solvents called amines (more properly, alkanolamines) are a family of organic compounds that are derivatives of alkanols (commonly called the alcohols group) that contain an "amino" (NH_2) group in their chemical structures. Because of this complexity, there are multiple classifications of amines, each of which has different characteristics relevant to CO_2 capture.[24] For example, MEA reacts strongly with acid gases like CO_2 and has a fast reaction time and an ability to remove high percentages of CO_2, even at the low CO_2 concentrations found in flue gas streams. Other properties of MEA, however, are undesirable, such as its high corrosivity and regeneration energy requirement. Various research groups are involved in synthesizing and testing a variety of amine mixtures and "designer" amines to achieve a more desirable set of overall properties for use in CO_2 capture systems. One major focus is on lowering the energy required for solvent regeneration, which has a major impact on process costs. Often, however, there are tradeoffs to consider. For example, the energy required for regeneration is typically related to the driving forces for achieving high capture capacities. Thus, reducing the regeneration energy can lower the driving force and thereby increase the amount of solvent and size of absorber needed to capture a given amount of CO_2—thus increasing the capital cost. A higher cost of manufacturing

[24] A. L. Kohl and R. B. Nielsen, *Gas Purification*, 5th ed. (Houston, TX: Gulf Publishing Company, 1997).

a new solvent also may detract from its benefits. Pilot plant projects are acquiring the data needed to assess such tradeoffs and optimize the overall process.

Ammonia-Based Capture Processes

A 2005 study by DOE/NETL found that post-combustion CO_2 capture using ammonia appeared very promising. It suggested that if a number of engineering challenges could be overcome, the overall cost of an ammonia-based system would be substantially less than an amine-based system for CO_2 capture. Since ammonia potentially could capture multiple pollutants simultaneously (including CO_2, SO_2, NO_x, and Hg), the overall plant cost could be reduced even further.[25]

Ammonia-based systems are attractive in part because ammonia is inexpensive, but also because an ammonia-based process potentially could operate with a fraction of the energy penalty of amines. Less compressor power also would be required, since CO_2 can be regenerated at higher pressure. These considerations led to early estimates that the overall energy penalty of an ammonia-based system could be reduced to about half that of a conventional amine system—claims not substantiated in subsequent testing. Ammonia also has a higher volatility than MEA and thus is more easily released into the flue gas stream during the absorption step (a process called "ammonia slip"). Controlling ammonia slip to acceptable levels is one of the major engineering challenges, since a need for subsequent cleanup would add considerably to the cost.[26] The development of ammonia-based capture technology has advanced to the pilot plant stage, with the intent of soon scaling up to commercial sizes. The two major companies involved in ammonia-based capture, a description of the pilot plants they have constructed, and the announced plans for this technology are described below.

The Alstom Chilled Ammonia Process

In the chilled ammonia process being developed by Alstom, the flue gas and CO_2 absorber are cooled to about 20°C (68°F), a temperature that prevents large amounts of ammonia slip from exiting the absorber with the cleaned flue gas stream. In the absorber, ammonium carbonate is used to capture the CO_2. As with amine system designs, the CO_2-"rich" stream is then sent to a stripper column where heat is added (using steam extracted from the power plant steam turbine) to strip CO_2 from the solution. This leaves a nearly pure CO_2 stream that can be cleaned, dried, and compressed for transport to a geological storage site. The CO_2-"lean" stream is then recirculated back to the absorber (**Figure 15**).

The energy required to regenerate the ammonia-based solvent is believed to be much smaller than for amine systems, which would considerably reduce the overall process cost. However, there is also an important tradeoff between the energy required to cool the process and the additional equipment and energy costs of reducing ammonia slip to acceptable levels. Thus, the overall process design must be optimized to achieve the best performance at minimum cost. Since details of the Alstom process remain proprietary, rigorous cost and performance comparisons with other processes are currently unavailable.

---

[25] U.S. Department of Energy, *An Economic Scoping Study for CO₂ Capture Using Aqueous Ammonia*, prepared by J. P. Ciferno, P. DiPietro, and T. Tarka, National Energy Technology Laboratory, Pittsburgh, PA (2005); http://www.transactionsmagazine.com/ArgonneLabCommonSense.pdf.

[26] D. Figueroa et al., "Advances in CO₂ capture technology—The U.S. Department of Energy's Carbon Sequestration Program," *International Journal of Greenhouse Gas Control*, vol 2 (2008), pp. 9-20.

Figure 15. Schematic of the Chilled Ammonia Process for CO_2 Capture (left) and the 20 MW Pilot Plant at the AEP Mountaineer Station in West Virginia (right)

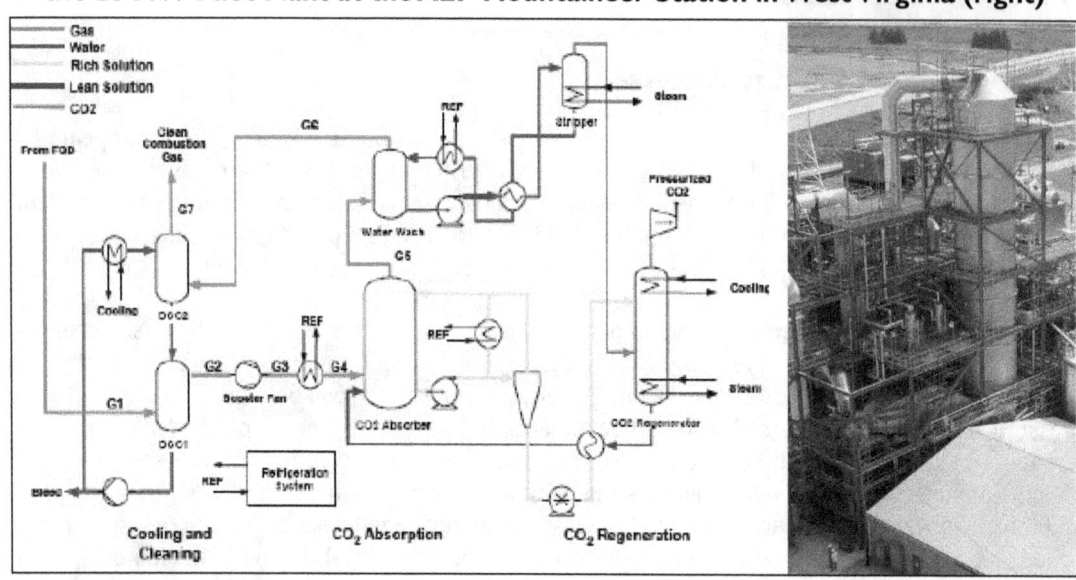

Source: Photo courtesy of AEP.

Alstom is currently operating two pilot plants using their chilled ammonia process—one in the United States and one in Norway (see **Table 7**). The most recent is the pilot plant at the American Electric Power (AEP) Mountaineer power station in West Virginia, a 1300 MW coal-fired plant, where a flue gas slip stream equivalent to about 20 MW has been fitted with the Alstom process (see **Figure 15**). This is the first successful integration of CO_2 capture, transport and geological sequestration at a coal-fired power plant. Data from this pilot plant will provide the basis for the proposed demonstration plant listed in **Table 6**.

The Powerspan ECO₂ Capture Process

Powerspan has developed a technology called the ECO process, which uses ammonia to capture SO_2 and NO_x from power plant flue gas streams in lieu of separate flue gas desulfurization and selective catalytic reduction systems. In 2005, Powerspan expanded the ECO process to also capture CO_2. This process, called ECO_2, is similar to the Alstom chilled ammonia process in that it also uses ammonium carbonate to capture CO_2, though the process operates at a higher temperature. Ammonium sulfate from the ECO process is used to control ammonia slip so that ammonia is not consumed in the process. Thus, while amine-based systems must severely limit exposure of the solvent to acid gases like SO_2 and NO_2 to prevent solvent loss and degradation, ammonia does not degrade in the presence of these gases; instead, it forms ammonium sulfate and nitrate, which have value as fertilizer by-products.[27] Powerspan is currently testing its ECO_2 process at a 1 MW pilot plant at First Energy's R. E. Burger plant, as indicated in **Table 7**.

---

[27] Clean Air Task Force, "Coal without Carbon."

Laboratory- or Bench-Scale Processes

A large number of new processes and materials for post-combustion CO_2 capture are currently at the laboratory- or bench-scale stage of development.[28] These can be grouped into three general categories: (1) liquid solvents (absorbents) that capture CO_2 via chemical or physical mechanisms; (2) solid adsorbents that capture CO_2 via physical mechanisms; and (3) membranes that selectively separate CO_2 from other gaseous species. Within each category, a number of approaches are being pursued, as summarized in **Table 8**.

Table 8. Post-Combustion Capture Approaches Being Developed at the Laboratory or Bench Scale

Liquid Solvents	Solid Adsorbents	Membranes
Advanced amines	Supported amines	Polymeric
Potassium carbonate	Carbon-based	Amine-doped
Advanced mixtures	Sodium carbonate	Integrated with absorption
Ionic liquids	Crystalline materials	Biomimetic-based

Source: Edward S. Rubin, Aaron Marks, Hari Mantripragada, Peter Versteeg, and John Kitchin, Carnegie Mellon University, Department of Engineering and Public Policy.

Each of the approaches in **Table 8** has some potential to reduce the cost and/or improve the efficiency of CO_2 capture relative to current commercial systems. At this early stage of development, however, it is difficult or impossible to reliably quantify the potential benefits or the likelihood of success in advancing to a commercial process. Indeed, at this stage, many of the approaches being investigated consist solely of a novel or advanced material that holds promise for CO_2 capture, but which remains to be developed into an engineered process that can properly be called a capture technology. Thus, even if a new material succeeds in capturing CO_2 more efficiently or with a lower energy penalty, substantial challenges remain in incorporating such materials into a viable and scalable technology that is more economical than current CO_2 capture systems.[29] While some of the approaches in **Table 8** may later advance to pilot-scale testing, others may not move past the bench scale. The sections below describe in greater detail the promise and challenges for each of these options.

Liquid Solvent-Based Approaches

Liquid solvents (typically a mixture of a base and water) selectively absorb CO_2 through direct contact between the chemical solvent and the flue gas stream. Regeneration of the solvent and release of CO_2 then takes place in a separate vessel (the regenerator) through a change of process conditions, such as a swing in temperature or pressure.

In general, the aim of solvent research is to identify or create new solvents or solvent mixtures that have more desirable characteristics than currently available solvents. Such properties include increases in CO_2 capture capacity, reaction rates, thermal stability, and oxidative stability, along with decreases in regeneration energy, corrosivity, viscosity, volatility, and chemical reactivity

---

[28] U.S. Department of Energy, *Proceedings of 1st Existing Plant Program Annual Meeting*, National Energy Technology Laboratory, Pittsburgh, PA, March 2008.

[29] Electric Power Research Institute (EPRI), *Post-Combustion CO2 Capture Technology Development*, Report No. 10117644, Technical Update, Palo Alto, CA, December 2009.

with flue gas impurities. All of these attributes tend to lower the cost of CO_2 capture compared to current solvents.

Unfortunately, most real solvents exhibit a combination of desirable and undesirable properties. Laboratory- and bench-scale research thus seeks new solvents that yield a more optimal blend of properties. **Table 9** summarizes the main advantages and challenges associated with liquid solvent-based approaches to post-combustion CO_2 capture.

Table 9. Technical Advantages and Challenges for Post-Combustion Solvents

Description	Advantages	Challenges
Solvent reacts reversibly with CO_2, often forming a salt. The solvent is regenerated by heating (temperature swing), which reverses the absorption reaction (normally exothermic). Solvent is often alkaline.	Chemical solvents provide fast kinetics to allow capture from streams with low CO_2 partial pressure. Wet scrubbing allows good heat integration and ease of heat management (useful for exothermic absorption reactions).	The large amount of steam required for solvent regeneration de-rates the power plant significantly. Energy required to heat, cool, and pump non-reactive carrier liquid (usually water) is often significant. Vacuum stripping can reduce regeneration steam requirements but is expensive; bad economy of scale. Multiple stages and recycle stream may be required.

Source: U.S. Department of Energy, *DOE/NETL Carbon Dioxide Capture R&D Annual Technology Update*, Draft, National Energy Technology Laboratory, Pittsburgh, PA, April 2010; hereafter "DOE, Carbon Dioxide Capture."

Examples of promising solvents include new amine formulations, carbonates, certain blends of amines and carbonates, and ionic liquids. For example, a promising new amines now receiving attention is piperazine. This solvent, currently being studied at the University of Texas and elsewhere, has been shown to have faster kinetics, lower thermal degradation and lower regeneration energy requirements than MEA in experiments thus far.[30] Further characterization studies are in progress.

Potassium carbonate solvents, which have been used successfully in other gas purification applications, are now being investigated for bulk CO_2 capture from flue gases.[31] Potassium carbonate absorbs CO_2 through a relatively low-energy reaction, but the process is slow. Researchers are attempting to speed up absorption by blending potassium carbonate with various amines, with promising results.[32] Modeling of piperazine-promoted blends, for example, has suggested that due to improved kinetics and low regeneration energy requirements, such systems could have smaller equipment sizes and be less energy intensive than MEA-based systems.[33]

---

[30] Rochelle, "Amine Scrubbing"; S. A. Freeman, J. Davis, and G. T. Rochelle, "Degradation of aqueous piperazine in carbon dioxide capture," *International Journal of Greenhouse Gas Control*, 2010.

[31] D. G. Chapel, C. L. Mariz, and J. Ernest, "Recovery of CO_2 from Flue Gases: Commercial Trends," Proc. *Canadian Society of Chemical Engineers Annual Meeting October 4-6, 1999*, Saskatoon, Saskatchewan, Canada. D. Wappel et al., "The Effect of SO_2 on CO_2 Absorption in an Aqueous Potassium Carbonate Solvent," *Energy Procedia*, vol. 1, no. 1 (2009), pp. 125-131. H. Knuutila, H. F. Svendsen, and O. Juliussen, "Kinetics of Carbonate-Based CO_2 Capture Systems," *Energy Procedia*, vol. 1 (2009), pp. 1011-1018.

[32] DOE, "Carbon Dioxide Capture"; J. T. Cullinane and G. T. Rochelle, "Carbon Dioxide Absorption with Aqueous Potassium Carbonate Promoted by Piperazine," *Chemical Engineering Science*, vol. 59 (2004), pp. 3619-3630.

[33] J. Oexmann, C. Hensel, and A. Kather, "Post-combustion CO_2-capture from Coal-fired Power Plants: Preliminary Evaluation of an Integrated Chemical Absorption Process with Piperazine-promoted Potassium Carbonate," (continued...)

Ionic liquids are liquid salts with low vapor pressure (hence, low flue gas losses) that potentially can absorb CO_2 at high temperatures with relatively low regeneration energy requirements.[34] Researchers at the University of Notre Dame have shown that ionic liquids can capture SO_2 as well as CO_2, leading to the possibility that they can be used in a multi-pollutant capture system.[35]

In a separate line of investigation, Georgia Tech Research Corporation is developing a class of solvents called reversible ionic liquids that chemically react with CO_2 to make other ionic liquids that further absorb CO_2.[36] One challenge for ionic liquids is that they can become highly viscous when absorbing CO_2, thus increasing the energy required for solvent pumping and the potential for mass transfer problems and operational difficulties in engineered processes.[37]

Solid Sorbent-Based Approaches

Solid sorbents capture (adsorb) CO_2 on their surfaces, as shown in **Figure 16**. They then release the CO_2 through a subsequent temperature or pressure change, thus regenerating the original sorbent. Solid sorbents have the potential for significant energy savings over liquid solvents, in part because they avoid the need for the large quantities of water that must be repeatedly heated and cooled to regenerate the solvent solution.[38] Sorbent materials also have lower heat capacity than solvents and thus require less regeneration energy to change their temperature.

A challenge, however, is how to efficiently get heat into and out of a solid sorbent material. More complicated solids handling equipment also is required compared to solvent solutions, which simply require pumps. In this regard, resistance to physical attrition and deterioration over time is another important property for most solid sorbent applications. Finally, it is not yet clear which of several different absorber designs that can utilize solid sorbents (e.g., fluidized beds, packed bed reactors, transport reactors, or other systems) will be most effective in reducing overall cost.

In general, the aim of solid sorbent research is to reduce the cost of CO_2 capture by designing durable sorbents with efficient materials handling schemes, increased CO_2 carrying capacity, lower regeneration energy requirements, faster reaction rates and minimum pressure drops.[39] The CO_2 carrying capacity is a key sorbent parameter that depends on the total microscopic surface area of the material. Researchers are thus attempting to identify and design sorbents with very high surface area for CO_2 capture.[40] The capture mechanism can be either a chemical or physical surface interaction. Solid sorbents that rely on chemical mechanisms are similar to liquid solvents. They include amines supported on the surface of other materials (called supported amines), as well as carbonates such as calcium carbonate (limestone) and sodium carbonate (soda ash). Sorbents that rely on physical surface interactions include materials such as activated carbon, zeolites, and metal organic frameworks (MOFs).

(...continued)

International Journal of Greenhouse Gas Control, vol. 2 (2008), pp. 539-552.

[34] GCCSI, "Strategic Analysis." DOE, "Carbon Dioxide Capture."

[35] Figueroa et al., "Advances CO_2 Capture."

[36] U.S. Department of Energy, *CO_2 Capture Technology Sheets*, report prepared by Leonardo Technologies Inc. for Existing Plants, Emissions and Capture Program, National Energy Technology Laboratory, Pittsburgh, PA, 2009.

[37] Clean Air Task Force, "Coal without Carbon."

[38] Figueroa et al., "Advances CO_2 Capture"; U.S. Department of Energy, "Carbon Dioxide Capture from Flue Gas using Dry Regenerable Sorbents," Project Facts, National Energy Technology Laboratory, Pittsburgh, PA, 2008.

[39] DOE, "Carbon Dioxide Capture."

[40] GCCSI, "Strategic Analysis."

Figure 16. Schematic of CO_2 Adsorption on the Surfaces of a Solid Sorbent

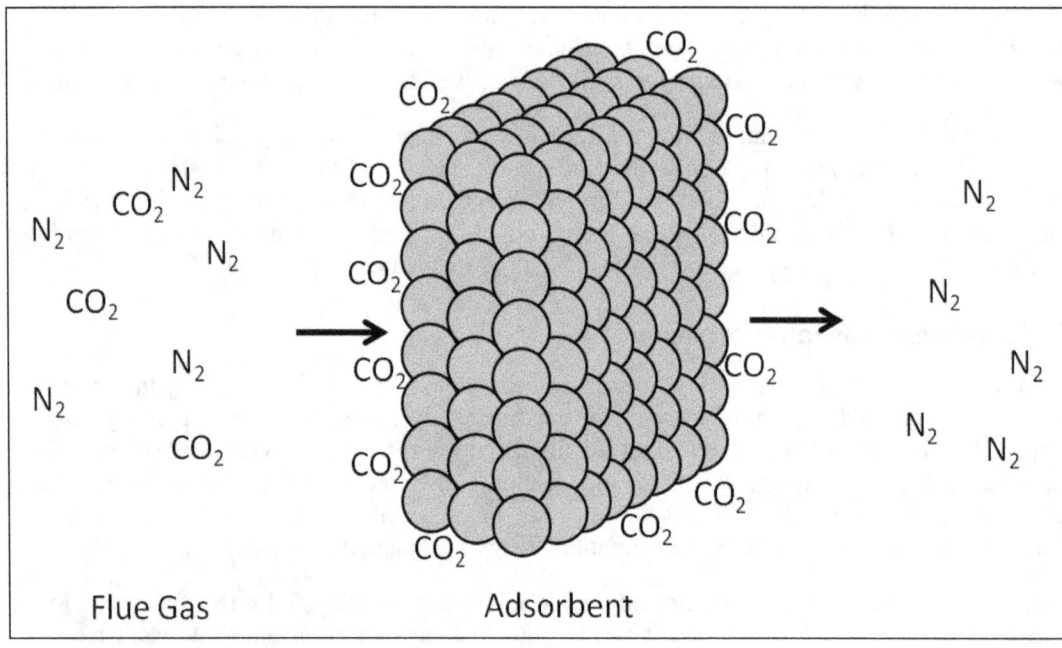

Source: Edward S. Rubin, Aaron Marks, Hari Mantripragada, Peter Versteeg, and John Kitchin, Carnegie Mellon University, Department of Engineering and Public Policy.

Notes: The simplified flue gas composition is represented as a mixture of CO_2 and nitrogen (N_2), the principal flue gas constituent.

Supported amines share the benefits of liquid amine solvents but require less energy to regenerate because there is no water solution.[41] The amine sorbent can be physically supported by a number of different materials, including relatively inexpensive activated carbon.[42] Such sorbents have been shown to have high CO_2 carrying capacities compared to other solid sorbents.[43] Current research is focused on issues of thermal stability and fouling, as these sorbents have a tendency to break down over time and degrade in the presence of SO_2.[44]

Sodium carbonate-based sorbents also have been recognized for their CO_2 capture potential,[45] although their performance is degraded by contaminants in flue gas.[46] Among the promising activities in this field is a CO_2 capture system using a sodium carbonate-based sorbent for use at coal or gas-fired power plants.[47]

---

[41] M. L. Gray et al., "Performance of Immobilized Tertiary Amine Solid Sorbents for the Capture of Carbon Dioxide," *International Journal of Greenhouse Gas Control*, vol. 2 (2008), pp. 3-8.

[42] M. G. Plaza et al., "CO_2 Capture by Adsorption with Nitrogen Enriched Carbons," *Fuel*, vol. 86 (2007), pp. 2204-2212.

[43] S. Sjostrom and H. Krutka, "Evaluation of Solid Sorbents as a Retrofit Technology for CO2 Capture," *Fuel*, vol. 89 (2010), pp. 1298-1306.

[44] DOE, "Technology Sheets."

[45] Y. Liang et al., "Carbon Dioxide Capture Using Dry Sodium-Based Sorbent," *Energy & Fuels*, vol. 18 (2004), pp. 569-575.

[46] GCCSI, "Strategic Analysis."

[47] Figueroa et al., "Advances CO_2 Capture."

Carbon-based adsorbents such as activated carbon and charcoal also are attractive because they are relatively inexpensive and have large surface areas that can readily adsorb CO_2. Researchers at the University of Wyoming, for example, claim that their Carbon Filter Process potentially can capture 90% of flue gas CO_2 and regenerate it with at least 90% CO_2 purity at a lower cost than amine-based processes.[48] They also can provide a support material for amines or other solid sorbents.

Metal organic frameworks and zeolites are crystalline sorbents that are also receiving attention for post-combustion CO_2 capture. MOFs consist of a matrix structure of metallic and organic molecules that contain void spaces that can potentially be used to absorb large amounts of CO_2 with low regeneration energy requirements and cost. Zeolites are porous alumino-silicate materials that have high selectivity, but low carrying capacity for CO_2 and are subject to performance degradation in the presence of water.[49] Researchers at the University of Akron are investigating an approach combining zeolites with amines to improve overall performance.[50]

Table 10 summarizes the key advantages and challenges of solid sorbent-based approaches to post-combustion CO_2 capture. Although such systems have the potential to offer better performance than current amine systems, the need to handle large amounts of solids tends to make this approach more complex and more difficult to scale up than an equivalent liquid solvent system. Sorbents also must have high selectivity for CO_2 and be relatively insensitivity to trace impurities in the flue gas. Because CO_2 bonding to sorbents is not as strong as with chemical interactions, multiple contacting stages also may be required to achieve high CO_2 capture efficiencies, which would increase process costs.[51] Current R&D programs are attempting to address these challenges.

Table 10. Technical Advantages and Challenges for Solid Sorbent Approaches to Post-Combustion CO_2 Capture

Description	Advantages	Challenges
When sorbent pellets are contacted with flue gas, CO_2 is absorbed onto chemically reactive sites on the pellet. Pellets are then regenerated by a temperature swing, which reverses the absorption reaction.	Chemical sites provide large capacities and fast kinetics, enabling capture from streams with low CO_2 partial pressure. Higher capacities on a per mass or volume basis than similar wet-scrubbing chemicals. Lower heating requirements than wet-scrubbing in many cases (CO_2 and heat capacity dependent).	Heat required to reverse chemical reaction (although generally less than for wet-scrubbing). Heat management in solid systems is difficult. This can limit capacity and/or create operational issues for exothermic absorption reactions. Pressure drop can be large in flue gas applications. Sorbent attrition may be high.

Source: DOE, "Carbon Dioxide Capture."

---

[48] M. Radosz et al., "Flue-Gas Carbon Capture on Carbonaceous Sorbents: Toward a Low-Cost Multifunctional Carbon Filter for 'Green' Energy Producers," *Industrial & Engineering Chemistry Research*, vol. 47 (2008), pp. 3784-3794; EPRI, 'Post-Combustion.'

[49] Clean Air Task Force, "Coal without Carbon."

[50] Figueroa et al., "Advances CO_2 Capture."

[51] Clean Air Task Force, "Coal without Carbon"; GCCSI, "Strategic Analysis."

Membrane-Based Approaches

Membranes are porous materials that can be used to selectively separate CO_2 from other components of a gas stream. They effectively act as a filter, allowing only CO_2 to pass through the material. The driving force for this separation process is a pressure differential across a membrane, which can be created either by compressing the gas on one side of the material or by creating a vacuum on the opposite side.

Membranes have been used for gas purification in a number of industrial applications since the 1980s.[52] Two important physical parameters of a membrane are its selectivity and permeability. Selectivity reflects the extent to which a membrane allows some molecules to be transported across the material, but not others. For post-combustion CO_2 capture, the selectivity to CO_2 over N_2 (the main constituent of flue gas) determines the purity of the captured CO_2 stream. The permeability of a membrane reflects the amount of a given substance that can be transported for a given pressure difference.[53] This determines the membrane surface area needed to separate and capture a given amount of CO_2.

Among the current laboratory- and bench-scale developments in this area, researchers at the University of Mexico are attempting to incorporate amine functional groups into membrane materials—a development that could help raise the selectivity of CO_2.[54] Another active research area is gas absorption membranes.[55] Here, CO_2-laden flue gases contact one side of a membrane while a liquid solvent (such as an amine-based solvent) contacts the other side. As CO_2 and other gases pass through the membrane, the CO_2 is selectively absorbed by the liquid solvent.[56] This approach holds potential for better performance than conventional absorber and stripper configurations.[57]

Yet another approach employs membranes with biomimetric components, seeking to employ processes found in nature. One such process uses the enzyme carbonic anhydrase, which facilitates the transport of CO_2 in the respiratory system of mammals.[58] One effort to exploit this process is a liquid membrane system catalyzed by carbonic anhydrase being developed by Carbozyme Inc.[59] While preliminary results show potential for significant decreases in energy penalty and cost compared to amine-based systems, the significant challenges that remain include the problems of membrane fouling and scale-up to power plant applications.

Table 11 summarizes the potential benefits and technical challenges of membrane-based technologies for post-combustion CO_2 capture. By most accounts, membranes today do not have the selectivity needed to be economically competitive with amine-based post-combustion CO_2

---

[52] J. Kotowicz, T. Chmielniak, and K. Janusz-Szymańska, "The Influence of Membrane CO_2 Separation on the Efficiency of a Coal-fired Power Plant," *Energy*, vol. 35 (2010), pp. 841-850. E. Favre, R. Bounaceur, and D. Roizard, "Biogas, Membranes and Carbon Dioxide Capture," *Journal of Membrane Science*, vol. 328, no. 1-2 (2009), pp. 11-14.

[53] L. Zhao et al., "Multi-stage Gas Separation Membrane Processes used in Post-Combustion Capture: Energetic and Economic Analyses," *Journal of Membrane Science*, Article in Press, pp. 1-13.

[54] Figueroa et al., "Advances CO_2 Capture"; Clean Air Task Force, "Coal without Carbon."

[55] EPRI, "Post-Combustion."

[56] GCCSI, "Strategic Analysis."

[57] Clean Air Task Force, "Coal without Carbon."

[58] E. Hand, "The Power Player," *Nature*, vol. 462 (2009), pp. 978-983.

[59] Figueroa et al., "Advances CO_2 Capture."

---

capture.[60] Additional challenges include the need for large surface areas to process power plant flue gases, limited temperature ranges for operation, low tolerance to flue gas impurities (or requirements for additional equipment to remove those impurities) and high parasitic energy requirements to create a pressure differential across the membrane.[61]

Table 11. Technical Advantages and Challenges for Membrane-Based Approaches to Post-Combustion CO_2 Capture

Description	Advantages	Challenges
Uses permeable or semi-permeable materials that allow for the selective transport and separation of CO_2 from flue gas.	No steam load. No chemicals needed.	Membranes tend to be more suitable for high-pressure processes such as IGCC. Tradeoff between recovery rate and product purity (difficulty to meet both at same time). Requires high selectivity (due to CO_2 concentration and low pressure ratio). Good pre-treatment. Poor economies of scale. Multiple stages and recycle streams may be required.

Source: DOE, "Carbon Dioxide Capture."

Despite these issues, there are strong proponents of membranes for post-combustion CO_2 capture. For example, Favre (2007) asserts that many of the challenges for membrane technology are amenable to engineering solutions. He also notes that membranes could be more competitive with amines in applications with higher CO_2 concentrations, such as in the cement and steel industries. A power plant boiler fired by oxygen-enriched air also would increase the CO_2 concentration of the flue gas, making membrane-based separation more competitive.[62]

Conceptual Design Stage

This stage of development typically involves engineering analyses or computer-based modeling studies of novel capture technology concepts or systems whose fundamental principles are usually well understood, but that lack the experimental data needed to test or verify the merits of the idea. This section briefly discusses three classes of novel but untested approaches to carbon capture: novel sorbents, hybrid systems, and novel regeneration methods.

Novel Sorbents

A number of research groups are investigating the development of ultra-high surface area porous materials for CO_2 capture. These materials are known as metal organic frameworks (MOFs, discussed earlier), zeolytic imidizolate frameworks, and porous organic polymers. These

---

[60] Clean Air Task Force, "Coal without Carbon."

[61] C. E. Powell and G. G. Qiao, "Polymeric CO_2/N_2 Gas Separation Membranes for the Capture of Carbon Dioxide from Power Plant Flue Gases," *Journal of Membrane Science*, vol. 279 (2006), pp. 1-49.

[62] Favre, "Biogas, Membranes."

materials have pore sizes, surface areas, and chemistries that are highly "tunable," meaning that molecules can, in principle, be designed and fabricated by chemists and materials scientists to maximize CO_2 capture performance. Because CO_2 capture research in this area is relatively new, very little work has yet been done to assess these materials under realistic capture conditions or to incorporate them into workable capture technologies.

Hybrid Capture Systems

Hybrid approaches to new solvents and sorbents attempt to combine the best features of two or more components to mitigate the undesirable properties of one component. For example, a typical problem with some CO_2 capture solvents is that they become highly viscous when interacting with CO_2. Hybrid approaches to solving this problem are to support the solvent on either a membrane or a solid sorbent. In these cases, viscosity is no longer an issue since no liquids are flowing.

For solid sorbents, one of the key problems is how to get heat into the sorbent during regeneration, since heat transfer in gas-solid systems is not as efficient as in liquid systems. One proposed solution is to immobilize the sorbents on a membrane or other solid support material that allows heat to be transferred more efficiently between two solids in direct contact.

Some examples of these hybrid approaches have advanced to the laboratory or bench scale, while others are being studied at the concept stage. It is uncertain, however, how the cost of these systems will compare to that of a single-component system whose active capture agent is now "diluted" by the other component. In general, one expects that the capital cost will be higher for a hybrid system, so its CO_2 capture performance must be substantially improved to offset the cost.

Novel Regeneration Methods

The two most common ways of regenerating CO_2 capture solvents or sorbents are application of heat (temperature swing) or a vacuum (pressure swing), both of which are energy-intensive and costly. Researchers are examining alternative approaches that could be more efficient and less costly.

One alternative (and theoretically more efficient) approach to regeneration is based in electrochemistry. A flow of electrons (electricity) is used to facilitate both the capture and regeneration steps. Of the several concepts that have been studied, the most promising applies electrochemistry to carbonate materials to make separate acid and base solutions (so-called pH swing systems), with one solution used as a solvent to capture CO_2 and the other used to regenerate the solvent.[63] This technology is similar to a fuel cell in that it requires electrodes and specialized membranes to selectively separate particular species, such as protons and hydroxide ions. **Figure 17** illustrates one of the conceptual designs.

---

[63] M. D. Eisaman, "CO_2 concentration using bipolar membrane electrodialysis," Proc. *Gordon Research Conference on Electrochemistry*, 2010.

Figure 17. Schematic of a Process Concept Using Electrodialysis to Capture and Regenerate CO₂, While Generating Hydrogen and Oxygen as By-Products

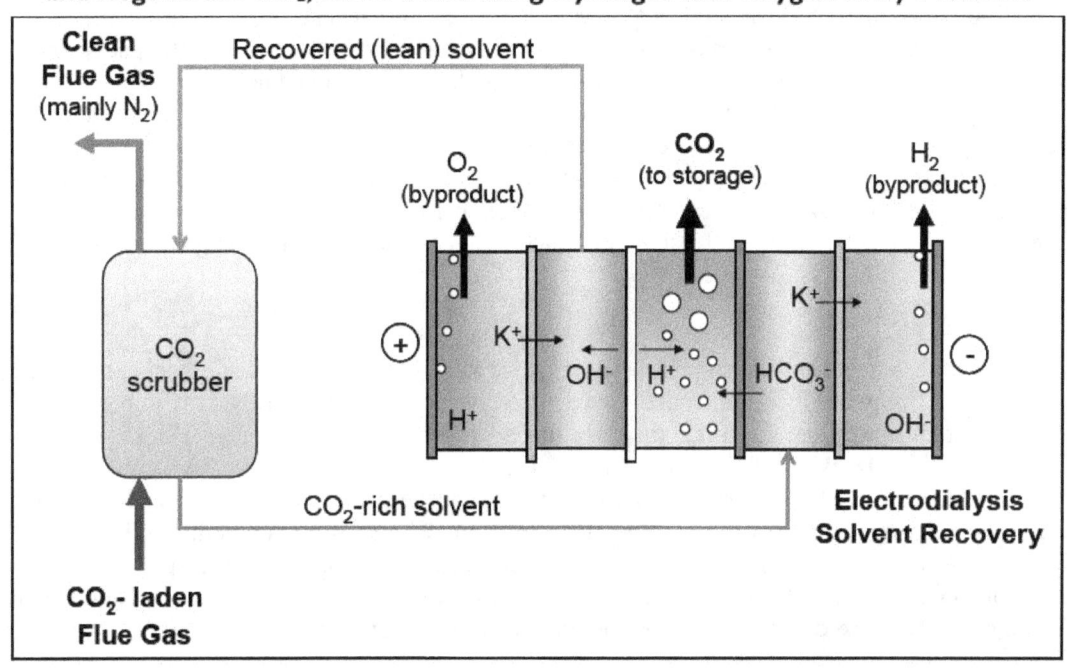

Source: Edward S. Rubin, Aaron Marks, Hari Mantripragada, Peter Versteeg, and John Kitchin, Carnegie Mellon University, Department of Engineering and Public Policy.

There are two variations of the pH swing concept, electrolysis and electrodialysis. The energy required for electrolysis is high and similar to that required for electrolysis of water. However, besides capturing CO_2 the process also generates hydrogen and oxygen, which have additional economic value. Electrodialysis is a more efficient process, but no valuable gases such as hydrogen are produced. Electrodialysis has been used commercially to desalinate water, but is only just being studied for application to CO_2 capture.[64]

A third electrochemical approach employs membranes to separate gases such as hydrogen, oxygen and CO_2. This approach is theoretically the most efficient, but high efficiencies have not been obtained in practice due to the limitations of existing materials.[65] While the fundamentals of electrochemical approaches to CO_2 capture have been proven at the bench scale, complete process designs are still only conceptual at this time.

Other concepts for regenerating CO_2 sorbents or solvents employ photochemical processes or electromagnetic radiation such as microwave heating.[66] At this point, however, it is unlikely that such approaches will soon (if ever) move out of the conceptual stage because of either technical or economic limitations.

---

[64] Advanced Research Projects Agency—Energy, "Energy Efficient Capture of CO_2 from Coal Flue Gas," at http://arpae.energy.gov/LinkClick.aspx?fileticket=CoktKdXJd6U%3d&tabid=90.

[65] H. W. Pennline et al., "Separation of CO_2 from flue gas using electrochemical cells," *Fuel*, vol. 89 (2010), pp. 1307-1314.

[66] Advanced Research Projects Agency—Energy, "ARPA-E's 37 Projects Selected from Funding Opportunity Announcement #1," http://arpa-e.energy.gov/LinkClick.aspx?fileticket=b-7jzmW97W0%3d&tabid=90.

System Studies

In addition to component-level studies of advanced CO_2 capture technologies, a variety of systems studies have been undertaken to analyze ways of improving the overall efficiency of power plants with CCS. One of the most promising methods is improved heat integration between the power plant and the CO_2 capture unit.[67] As noted in "Chapter 3: Overview of CO_2 Capture Technologies," measures that increase plant efficiency can also reduce the cost of CO_2 capture, provided that they do not introduce new costs that offset the efficiency benefits. Implementation of such measures must await the construction of large-scale demonstration plants or fully integrated pilot plants where the feasibility of such designs can be evaluated in greater detail.

Conclusion

This chapter has reviewed and summarized the major R&D activities aimed at reducing the cost of post-combustion CO_2 capture. While such activities have increased substantially in recent years, most current efforts are still at the early stages of technology development. This is seen clearly in **Figure 18**, which shows the results of a study by the Electric Power Research Institute that reviewed over 100 active projects in this field and ranked them on the TRL scale described earlier in "Chapter 4: Stages of Technology Development."[68] That study found that all but a few of the post-combustion capture projects were between TRLs 1 and 5, which corresponds to the conceptual design and laboratory-bench scale categories used in this report. Only a small number of projects were ranked at TRL 6, corresponding to the pilot plant stage in this report.

Figure 18. Technical Readiness Levels (TRLs) of Projects Developing Post-Combustion Capture Technologies Using Different Approaches

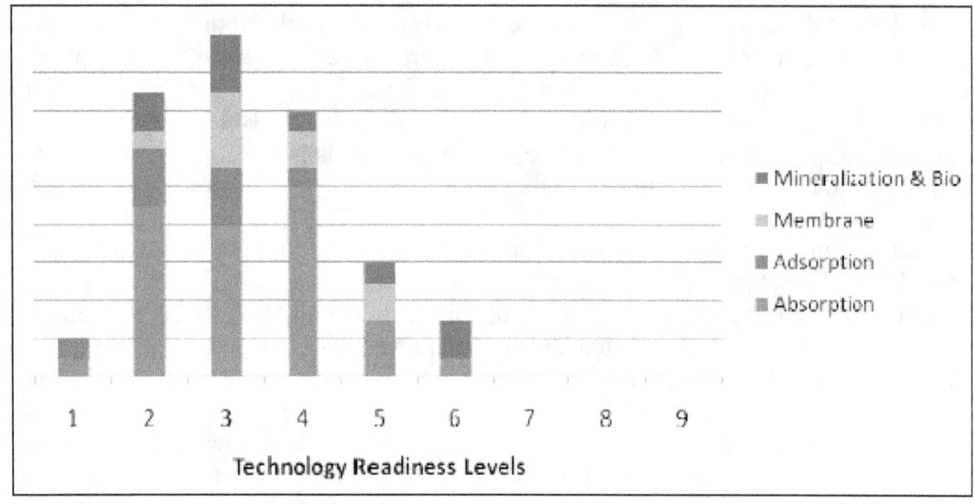

Source: Bhown and Freeman, "Assessment Post-Combustion."

Notes: The y-axis is not scaled explicitly but corresponds to the relative number of processes of a given type. Also, the approach labeled "Mineralization & Bio" is considered in the present report to be a sequestration method rather than a post-combustion capture method since it typically requires a stream of concentrated CO2 that has already been captured.

---

[67] David C. Thomas, ed., *The CO₂ Capture and Storage Project (CCP) for Carbon Dioxide Storage in Deep Geologic Formations for Climate Change Mitigation, Volume 1—Capture and Separation of Carbon Dioxide from Combustion Sources* (London: Elsevier, 2004); Rubin, 'Cost and Performance.'

[68] Bhown and Freeman, "Assessment Post-Combustion."

The EPRI study also shows that most of the new processes under development employ absorption methods (i.e., solvents) for post-combustion capture of CO_2. Fewer new processes and concepts utilize membranes or solid sorbents (adsorption) for CO_2 capture—a reflection of the greater challenges facing those approaches.

Key questions that remain are: What are the prospects for any of these projects to result in a viable new process for CO_2 capture? How much improvement in performance or reduction in cost can be expected relative to current or near-term options? How long will it take to see these improvements? Such questions are addressed later in "Chapter 7: Status of Oxy-Combustion Capture" and "Chapter 8: Cost and Deployment Outlook for Advanced Capture Systems," following a status report on the two other major approaches to CO_2 capture.

Chapter 6: Status of Pre-Combustion Capture

Introduction

This chapter summarizes the status of current and emerging pre-combustion CO_2 capture technologies at various stages of development. Pre-combustion CO_2 capture can be used both in power plants and in other industrial processes where CO_2 separation is required, such as in synthetic fuels production. The more advanced capture systems include chemical solvents such as Selexol^{TM} and Rectisol®, which are used widely in natural gas and synthesis gas applications. Processes at the earliest stages of development employ novel methods such as solid sorbents or membranes for CO_2 capture. The chapter begins with a discussion of commercial processes and then describes technologies at the less advanced stages of development outlined in "Chapter 4: Stages of Technology Development."

As noted previously, carbon capture research and development programs throughout the world have expanded rapidly in recent years; thus any summary of "current" activities and projects soon grows out of date. For this reason, there is no attempt in this report to be comprehensive in the coverage of capture-related R&D activities. Rather, this report attempts to synthesize key findings from our own investigations and from the work of others who also track and report on the status of CO_2 capture technology developments. This report draws too upon a set of publicly available databases and CCS project status reports maintained by the U.S. Department of Energy (DOE), the International Energy Agency's Greenhouse Gas Control Programme (IEAGHG), the Massachusetts Institute of Technology carbon sequestration program (MIT), and the Global Carbon Capture and Storage Institute (GCCSI).

In each of the sections below, the objective is to summarize not only the current status of technological developments (as of March 2010), but also the key technical barriers that must be overcome to advance pre-combustion capture methods, along with the potential payoffs in terms of improved performance and/or reduced costs. Brief descriptions of new capture methods or processes not previously discussed in "Chapter 3: Overview of CO_2 Capture Technologies" also are provided.

Commercial Processes

Currently there are no commercial applications of pre-combustion CO_2 capture at electric power plants. However, the Selexol^{TM} and Rectisol® processes that would be used in an IGCC power plant are already widely used in other commercial applications, mainly for removing contaminants such as sulfur and nitrogen compounds from syngas mixtures, as well as for capturing CO_2 present in syngas. Two examples are cited here to illustrate the scale at which pre-combustion capture technologies are currently used commercially.

The Farmlands chemical plant in Coffeyville, Kansas, shown in **Figure 19**, uses the Selexol system to separate and capture CO_2 from a hydrogen-CO_2 gas mixture produced by the gasification of petroleum coke (petcoke) followed by a water-gas shift reactor—the same processes depicted earlier in **Figure 6** for an IGCC with pre-combustion CO_2 capture. At the Coffeyville plant, more than 93% of the CO_2 is captured, amounting to about 0.2 million tons of

CO_2 per year.[69] A portion of this CO_2 is used to manufacture urea and the remainder is vented to the atmosphere. The separated stream of nearly pure hydrogen is used to manufacture ammonia (rather than burned to generate electricity, as in an IGCC plant), with the ammonia subsequently used to produce fertilizers. This project has been in operation since 2000 and is similar to other industrial applications that use the Selexol process for CO_2 capture.

The Great Plains synfuels plant in North Dakota operated by the Dakota Gasification Company, also shown in **Figure 19**, employs coal gasification to produce synthetic natural gas. In that process, the plant captures approximately 3 million tons/year of CO_2 using the methanol-based Rectisol process. Previously, the CO_2 was vented to the atmosphere. Now the CO_2 is compressed and transported via a 205-mile pipeline to a Canadian oil field, where it is used for EOR and sequestered in the depleted oil reservoir.

Figure 19.A Pre-Combustion CO_2 Capture System Is Used to Produce Hydrogen from Gasified Petcoke at the Farmlands Plant in Kansas (left) and Synthetic Natural Gas from Coal at the Dakota Gasification Plant in North Dakota (right)

Source: Photos courtesy of UOP and IPCC.

These two examples illustrate current commercial applications of pre-combustion CO_2 capture technologies that would be employed at gasification-based power plants. The choice of solvent or process would depend on the conditions of a particular project or application. The following section discusses current plans for full-scale demonstrations of pre-combustion capture at power plants.

Full-Scale Demonstration Plants

As with post-combustion capture, to date there have been no full-scale demonstrations of pre-combustion CO_2 capture at an IGCC power plant, although a number of full-scale projects have been announced and one (in China) is currently under construction. Several other previously announced IGCC-CCS projects in different parts of the world have been canceled or delayed in

---

[69] D. Heaven et al., "Synthesis Gas Purification in Gasification to Ammonia/Urea Plants," *Gasification Technologies Conference, October, 2004, Washington, DC*, Gasification Technologies Council, San Francisco, CA.

recent years, including the highly publicized FutureGen project proposed for construction in Mattoon, Illinois. The fate of this jointly funded government-industry venture is still being negotiated as of this writing.[70] Nevertheless, it appears reasonable that at least some of the large-scale projects currently planned for pre-combustion CO_2 capture in the United States and other countries will indeed materialize over the next several years, with costs shared between the public and private sectors.

Table 12 lists the features and locations of major announced demonstration of pre-combustion CO_2 capture. They include fuels production plants and IGCC power plants.

Table 12. Planned Demonstration Projects with Full-Scale Pre-Combustion Capture

Project Name and Location	Plant and Fuel Type	Expected Year of Startup	Plant Size or Capacity	CO_2 Capture System	Annual CO_2 Captured (10^6 tonnes)
United States					
Baard Energy Clean Fuels (Wellsville, Ohio)	Coal+biomass to liquids	2013	53,000 barrels/day	Rectisol	N/A
DKRW Energy (Medicine Bow, Wyoming)	Coal to liquids	2014	20,000 barrels/day	Selexol	N/A
Summit Power (Penwell, Texas)	Coal IGCC	2014	400 MW$_g$	Selexol	3.0
Taylorville Energy Center (Taylorville, Illinois)	Coal IGCC	2014	602 MW	N/A	N/A
Mississippi Power, Kemper County IGCC (Mississippi)	Lignite IGCC	2014	584 MW	N/A	N/A
Wallula IGCC (Walla Walla County, Washington)	Coal IGCC	2014	600-700 MW	N/A	N/A
Hydrogen Energy (Kern County, California)	Petcoke IGCC	2015	250 MW	N/A	N/A
Southern California Edison IGCC (Utah)	Coal IGCC	2017	500 MW	Selexol	3.5
FutureGen Alliance (Mattoon, Illinois)[a]	Coal IGCC	>2012[a]	275 MW	N/A	N/A
Outside the United States					
GreenGen (Tianji Binhai, China)	Coal IGCC and poly-generation	2011 (stage I)	250 MW	N/A	N/A
Eston Grange IGCC (Teesside, UK)	Coal IGCC	2012	800 MW	N/A	5
Hartfield IGCC (Hartfield, UK)	Coal IGCC	2014	900 MW	Selexol	4.5
Genesee IGCC (Edmonton, Canada)	Coal IGCC	2015	270 MW	N/A	1.2
RWE Goldenbergwerk (Hurth, Germany)	Lignite IGCC	2015[b]	360 MW	N/A	2.3

---

[70] Air Products, "Air Products and EPRI Working Together on ITM Oxygen Technology for Use in Advanced Clean Power Generation Systems," at http://www.airproducts.com.

Project Name and Location	Plant and Fuel Type	Expected Year of Startup	Plant Size or Capacity	CO_2 Capture System	Annual CO_2 Captured (10^6 tonnes)
Kedzierzyn Zero Emission Power and Chemicals (Opole, Poland)	Coal-biomass IGCC and polygen	2015	309 MW	N/A	2.4
Nuon Magnum[c] (Eemshaven, Netherlands)	Multi-fuel IGCC	2015	1200 MW$_g$	N/A	N/A
ZeroGen (Rockhampton, Australia)	Coal IGCC	2015	530 MW$_g$	MHI	N/A
FuturGas (Kingston, Australia)	Lignite to liquids	2016	10,000 barrels/day	N/A	1.6

Sources: DOE, "NETL Carbon"; IEAGHG, "CO_2 Capture"; MIT, "Carbon Capture"; GCCSI, "Strategic Analysis."

Note: MW = net electrical megawatts output; MW$_g$ = gross electrical megawatts output.

a. Final decision still pending as of May 2010.

b. Depends on outcome of the Carbon Storage Law.

c. Depends on performance of the Buggenum pilot plant (see **Table 13**).

Most of the projects in **Table 12** would not begin operation until 2014 or later. In most cases the captured CO_2 would be sequestered in a depleted oil reservoir in conjunction with EOR. The percentage of CO_2 captured varies widely across the projects, from 50% to 90% of the carbon in the feedstock. **Table 12** shows that Selexol is the preferred technology for pre-combustion capture at projects that have announced their selection. However, for most of the projects listed the choice of solvent or capture technology is not yet known. China's GreenGen project is on track to become the first full-scale IGCC plant with CO_2 capture to become operational, with construction begun in 2009. Construction has not yet started on any of the other proposed projects.

Given the extensive commercial experience and scale of CO_2 capture in industrial processes with gas streams nearly identical to an IGCC plant, most of the large-scale projects in **Table 12** will serve to demonstrate other aspects of IGCC technology. In particular, the reliability of gasifier operations and the large-scale use of hydrogen to power the gas turbine following CO_2 capture are key technical issues that remain to be demonstrated in the electric utility environment. The plant startup schedules in **Table 12** indicate that it will be at least five years before significant operational data begins to accrue at most of the planned demonstration projects. As before, the possibility also remains that some of these planned projects may not materialize for economic or other reasons.

Pilot Plant Projects

In general there is relatively little current development of pre-combustion CO_2 capture at the pilot plant scale. However, two projects under construction at IGCC plants in Europe—Nuon's Buggenum plant in the Netherlands and Elcogas's Puertollano plant in Spain—are significant developments because they will be the first applications of CO_2 capture at operating IGCC facilities, albeit at a small scale treating only a portion of the syngas stream. These projects are expected to begin operation in the late 2010 and 2011 time frames, as shown in **Table 13**.

Table 13. Pilot Plant Projects for Pre-Combustion CO₂ Capture at IGCC Power Plants

Project Name and Location	Plant and Fuel Type	Expected Year of Startup	Plant Size or Capacity	CO₂ Capture System	Annual CO₂ Captured (10^6 tonnes)
Nuon Buggenum (Buggenum, Netherlands)	Coal and biomass IGCC	2010	N/A	Different physical and chemical solvents	0.010
Elcogas Puertollano (Puertollano, Spain)	Coal and petcoke IGCC	2011	14 MW_{th} (~5 MW)	Different commercial solvents	0.035

Sources: DOE, "NETL Carbon"; IEAGHG, "CO₂ Capture"; MIT, "Carbon Capture"; GCCSI, "Strategic Analysis."

Note: MW = net electrical megawatts output; MW_{th} = gross thermal energy input (megawatts).

The Nuon Buggenum project is aimed at testing pre-combustion CO_2 capture in order to better select, design, and optimize a capture system after some operating experience is gained. Both the water gas shift reactors and the CO_2 capture process will be optimized for their performance efficiency and different physical and chemical solvents will be tested. The main aim of this pilot plant is to gain operational experience that can be used at the much bigger Nuon Magnum IGCC power plant listed in **Table 12**.[71]

Laboratory- or Bench-Scale Developments

Although pre-combustion CO_2 capture has a lower energy penalty and lower cost than post-combustion capture processes performing similar duty, there is scope for further improvements that can reduce costs. With this aim, current research is focused mainly on improving the capture efficiency so that the size and cost of equipment can be lowered. Current research is focused on the same three approaches discussed in "Chapter 5: Status of Post-Combustion Capture" for post-combustion capture technologies, namely, liquid solvents, which separate CO_2 from a gas stream by absorption; solid sorbents, which separate CO_2 from by adsorption onto the solid surface; and membranes, which separate CO_2 by selective permeation through thin layers of solid materials.

Solvent-Based Capture Processes

As noted previously, current pre-combustion CO_2 capture systems employ solvents that selectively absorb CO_2 (and other acid gases) from a gas stream via the mechanism of physical absorption into the solvent. Physical absorption is characterized by weak binding forces between gas molecules and the solvent molecules. Research on physical solvents is aimed at improving the CO_2 carrying capacity and reducing the heat of absorption. Higher carrying capacity means that more CO_2 is captured in every pass through the absorption tower, thus lowering costs. Solvents with a low heat of absorption require less energy to strip CO_2 during the regeneration step, which also lowers cost. Of the two properties, the main focus is on improving the CO_2 carrying capacity, since the heat of absorption already is low for most physical solvents (which is also why a

---

[71] Nuon, "Towards 2nd Generation IGCC Plants, Nuon Magnum Multi-Fuel Power Plant," Proc. *Future of Coal and Biomass in a Carbon-Constrained World*, November 2, 2009, Fargo, ND.

pressure-swing method can be used to strip captured CO_2 from the solvent, unlike chemical solvents, where heat is needed).

The CO_2 carrying capacity of a solvent depends on a number of factors, including certain properties of the solvent, the partial pressure of CO_2 in the gas stream, and the temperature of the process. Usually, the carrying capacity increases at higher pressure and lower temperature. A practical problem with liquid solvents is their ability to corrode equipment materials. Any novel solvent must therefore have low corrosive properties. **Table 14** summarizes the advantages of physical solvents and the challenges in improving their properties.

Table 14. Key Advantages and Challenges of Physical Solvents for Pre-Combustion CO_2 Capture

Description	Advantages	Challenges
Solvent readily dissolves CO_2. Solubility is directly proportional to CO_2 partial pressure and inversely proportional to temperature, making physical solvents more applicable to low temperature, high pressure applications (cooled syngas). Regeneration normally occurs by pressure swing.	CO_2 recovery does not require heat to reverse a chemical reaction.	CO_2 pressure is lost during flash recovery.
	Common for same solvent to have high H_2S solubility, allowing for combined CO_2/H_2S removal.	Must cool synthesis gas for CO_2 capture, then heat and humidify again for firing in gas turbine.
	System concepts that recover CO_2 with some steam stripping rather than flashed, with delivery at a higher pressure, may optimize processes for power systems.	Low solubility can require circulating large volumes of solvent, which increases energy needs for pumping.
		Some H_2 may be lost with the captured CO_2.

Source: DOE, "Carbon Dioxide Capture."

Current research on new or improved solvents for pre-combustion capture seeks to develop solvents that allow CO_2 to be captured at higher pressures and temperatures. Currently, syngas from the gasifier must be cooled to near room temperature before entering the solvent-based CO_2 capture unit. After capture, the syngas must be reheated to prepare it for downstream processes. New solvents that can capture CO_2 at higher temperatures can therefore increase overall plant efficiency and potentially reduce the equipment needs and cost of CO_2 capture. In this context, ionic liquids, discussed earlier in "Chapter 5: Status of Post-Combustion Capture," are also being studied as potential solvents for CO_2 capture in pre-combustion applications.

As also noted in "Chapter 5: Status of Post-Combustion Capture," ionic liquids are salts that are liquid at room temperature. They have high CO_2 absorption potential and do not evaporate at temperatures as high as 250°C. In an IGCC system, this could allow separation of CO_2 without cooling the syngas, thereby reducing equipment size and cost. This is also one of the approaches being pursued to develop new physical absorption solvents for pre-combustion capture.[72]

Sorbent-Based Capture Processes

Solid sorbents are another class of material that potentially could be used for pre-combustion CO_2 capture as well as for post-combustion capture (see "Chapter 5: Status of Post-Combustion

---

[72] DOE, "Carbon Dioxide Capture."

Capture"). The primary advantage of solid sorbent systems over solvents in pre-combustion applications is their ability to operate at high temperatures. This avoids the additional equipment for syngas cooling, thus reducing cost. However, the handling of solids is generally more difficult than the handling of liquid-based systems. This offsets some of the advantages of solids and can be an important factor in the choice (and overall cost) between solvent and sorbent-based capture technology in large-scale applications.

Solid sorbent-based systems are used commercially today in a variety of applications, such as in hydrogen purification processes employing pressure swing adsorption. With some changes, that system has the scope to be adapted to capture CO_2. Lehigh University, RTI International, TDA Research, the University of North Dakota Energy & Environmental Research Center, and the URS Group are among the organizations currently working on development of solid sorbents.[73] The work is primarily focused on identifying the most promising sorbent materials and conducting bench-scale experiments. **Table 15** summarizes the key advantages and challenges of using solid sorbents for pre-combustion CO_2 capture.

Table 15. Key Advantages and Challenges of Solid Sorbents for Pre-Combustion CO_2 Capture

Description	Advantages	Challenges
When sorbent pellets are contacted with syngas, CO_2 is physically adsorbed onto sites and/or dissolves into the pore structure of the solid. Rate and capacity are directly proportional to CO_2 partial pressure, making these sorbents more applicable to high pressure applications. Regeneration normally occurs by pressure swing.	CO_2 recovery does not require heat to reverse a reaction. Common for H_2S to also have high solubility in the same sorbent, meaning CO_2 and H_2S capture can be combined. System concepts in which CO_2 is recovered with some steam stripping rather than flashed and delivered at a higher pressure may optimize processes for power systems.	Solids handling is more difficult than liquid-gas systems. CO_2 capture with sorbents is a novel concept though other gas purification processes use adsorption techniques.

Source: DOE, 'Carbon Dioxide Capture.'

Membrane-Based Capture Processes

As described in "Chapter 5: Status of Post-Combustion Capture," membrane-based capture processes operate by selectively allowing a gas to permeate through the membrane material. Membranes for CO_2 capture are made of micro-porous metallic, polymeric, or ceramic materials. For effective CO_2 capture in pre-combustion applications, they should not only have high permeability and selectivity to CO_2 but also be able to operate at the high pressures and temperatures characteristic of IGCC systems.

Figure 20 shows a schematic of a membrane separation process for CO_2 capture in an IGCC application, where CO_2 is preferentially separated from hydrogen in the gas stream following the water-gas shift and sulfur removal steps illustrated earlier in **Figure 6**. Because the separation is seldom perfect, several stages are typically needed to increase the purity of the separated components.

---

[73] DOE, 'Carbon Dioxide Capture.'

Figure 20. Schematic of Pre-Combustion CO₂ Capture Using a Membrane to Separate CO₂ and H₂ in the Gas Stream of an IGCC Power Plant

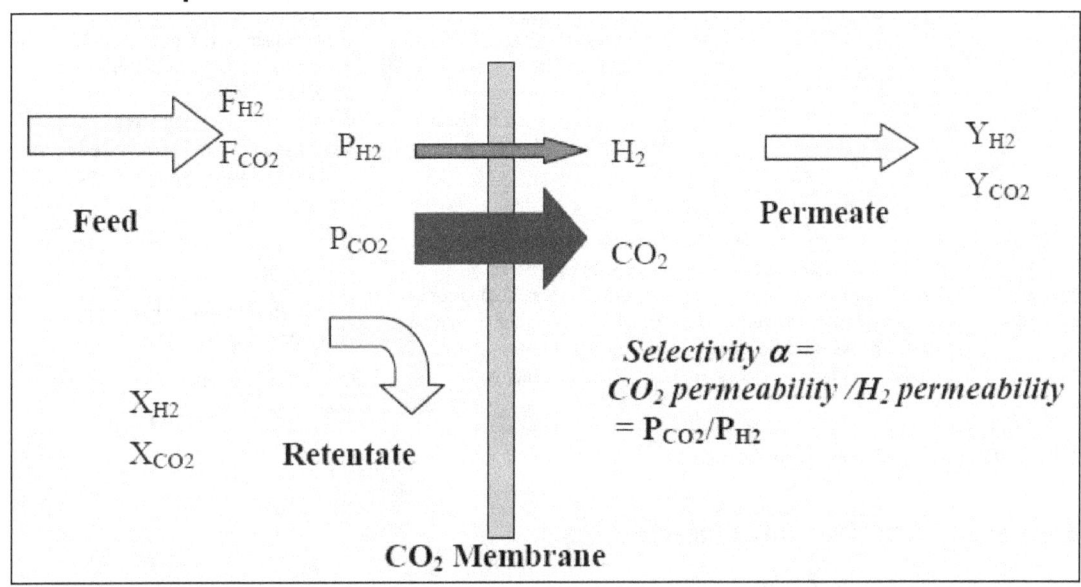

Source: DOE, "Carbon Dioxide Capture."

To date, membrane technology has been used commercially for gas purification and CO_2 removal in the production of hydrogen, but it has not been used specifically for pre-combustion CO_2 capture in IGCC plants or related industrial processes that require a high CO_2 recovery rate with high CO_2 purity. Applications to IGCC are of interest since the mixture of CO_2 and H_2 following the shift reactor is already at high pressure, unlike post-combustion applications, which require additional energy to create a pressure differential across the membrane.

Table 16 summarizes the key advantages and challenges of membrane separation systems for pre-combustion capture applications. Many of the challenges discussed earlier for post-combustion applications also apply here, such as limited temperature ranges for operation and low tolerance to impurities. Because of their modular nature and the need for relatively large surface areas, membrane systems again do not have the economies of scale with plant size found in other types of capture systems. Thus, they must have substantially superior performance and/or lower unit cost to compensate for this disadvantage. These are the major hurdles that current research is attempting to overcome.

Table 16. Key Advantages and Challenges of Membrane Separation Systems for Pre-Combustion CO₂ Capture

Membrane Type	Description	Advantages	Challenges
H_2-CO_2 membranes	A membrane material which selectively allows either H_2 or CO_2 to permeate through the material; potential use in gasification processes with streams of concentrated H_2 and CO_2.	H_2 or CO_2 Permeable Membrane: No steam load or chemical losses. H_2 Permeable Membrane Only: Can deliver CO_2 at high-pressure, greatly reducing compression costs.	Membrane separation of H_2 and CO_2 is more challenging than the difference in molecular weights implies. Due to decreasing partial pressure differentials, some H_2 will be lost with the CO_2.

Membrane Type	Description	Advantages	Challenges
		H_2 permeation can drive the CO shift reaction toward completion, potentially achieving the shift at lower cost/higher temperatures.	In H_2-selective membranes, H_2 compression is required and offsets the gains of delivering CO_2 at pressure. In CO_2 selective membranes, CO_2 is generated at low pressure, thus requiring added compression.
Membrane-Liquid Solvent Hybrids	Flue gas is contacted with a membrane and a solvent on the permeate side absorbs CO_2 and creates a partial pressure differential to draw CO_2 across the membrane.	The membrane shields the solvent from flue gas contaminants, reducing losses and allowing higher loading differentials between lean and rich solvent streams.	Capital cost associated with the membrane. Membranes may not keep out all unwanted contaminants. Does not address CO_2 compression costs.

Source: DOE, "Carbon Dioxide Capture."

Enhanced Water Gas Shift Reactors

As noted in "Chapter 3: Overview of CO_2 Capture Technologies," in an IGCC plant with CCS, the syngas exiting the gasifier is subjected to a water-gas shift (WGS) reaction to increase the concentration of CO_2 in the gas stream. This process is needed for efficient CO_2 capture in a subsequent step. It also provides additional hydrogen (H_2) for power generation. The WGS reaction between carbon monoxide (CO) in the syngas and steam (H_2O) that is added is:

$$CO + H_2O \Leftrightarrow CO_2 + H_2$$

The thermodynamics of chemical reactions dictates that the speed and efficiency of this reaction is limited by the presence of the reaction products (CO_2 and H_2) in the reactor vessel. Thus, to get high conversion efficiency of CO to CO_2, a catalyst is used and the WGS reaction is accomplished in two stages (and two vessels), with intermittent cooling of the syngas to help speed the reaction. This additional equipment and the associated energy penalty of the WGS process add to the cost of CO_2 capture.

To reduce this cost, researchers are developing sorbents and membranes that can be used within a WGS reactor so that the shift reaction occurs with simultaneous capture of CO_2. Thus, in a sorbent-enhanced water gas shift, the WGS catalyst is mixed with a CO_2 capture sorbent in a single reactor vessel. The sorbent removes CO_2 as soon as it is formed, which allows increased conversion of CO to CO_2. In this way, CO_2 capture is achieved simultaneously with an efficient WGS reaction, which can lower the overall capital cost of the system.[74] As with other sorbent-based capture schemes, however, the development of enhanced WGS reactors also requires a practical method of handling and regenerating the solid sorbent materials. This too is a topic of ongoing research.

A similar concept for simultaneous WGS and CO_2 capture employs a membrane reactor in which either CO_2 or H_2 is separated as soon as it is formed. Again, the removal of reaction products

---

[74] Van Dijk et al., "Performance of water-gas shift catalysts under sorption-enhanced water-gas shift conditions," *Energy Procedia*, vol. 1 (2009), pp. 639-646.

improves the effectiveness and speed of the WGS reaction. The possibility of using liquid solvents together with membranes also is being studied as a means of increasing the overall capture efficiency.[75]

Conceptual Design Stage

At the conceptual design stage, most of the work related to pre-combustion capture is focused on improving the efficiency of the overall power plant, which in turn lowers the cost of CO_2 capture (see "Chapter 3: Overview of CO_2 Capture Technologies"). Thus, improvements in all major IGCC system components—especially the air separation unit (ASU), gasifier and gas turbine—also are of interest for CO_2 capture. So too are studies of improved heat integration to reduce energy losses; advanced plant designs that integrate components such as the ASU and gas turbine air compressor; gasifier improvements that increase plant utilization; and advanced design concepts such as an IGCC system coupled with a solid oxide fuel cell. Examples of such studies can be found in several recent studies.[76]

Figure 21 shows an example of the cost reductions projected by the U.S. Department of Energy for conceptual designs of IGGC systems employing a variety of advanced technologies. These advances also would reduce the incremental cost of CO_2 capture. Substantial R&D efforts would be needed, however, to bring such designs to commercial reality.

Figure 21. Projected Cost Reductions for IGCC Systems Employing Advanced Technologies

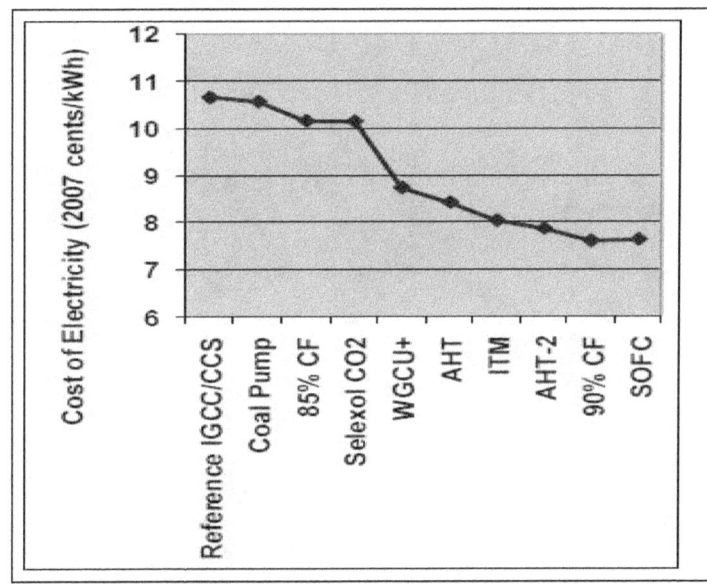

Source: Klara and Plunkett, "Potential Advanced."

Notes: These improvements also reduce the cost of CO_2 capture. Terms not defined previously: CF = capacity factor; WGCU = warm gas cleanup; AHT = advanced hydrogen-fired turbines (designs 1 and 2); ITM = ion transport membrane (for O_2 production); SOFC = solid oxide fuel cell (integrated with gasifier).

---

[75] DOE, "Carbon Dioxide Capture."

[76] Chen and Rubin, 'CO_2 Control Technology.' J. M. Klara and J. E. Plunkett, "The potential of advanced technologies to reduce carbon capture costs in future IGCC power plants," *International Journal of Greenhouse Gas Control*, Special Issue, Elsevier, March 2010.

Conclusion

This chapter has reviewed and summarized the major research and development activities aimed at reducing the cost of pre-combustion CO_2 capture. Many of these activities are similar in nature to those for post-combustion capture insofar as they involve the same basic concepts for new or improved capture processes. Improvements also are being sought in a variety of other IGCC plant components that also affect CO_2 capture costs, such as the air separation unit, gasifier, water-gas shift reactor, and gas turbines. At the conceptual level, advanced plant designs employing new plant integration concepts and advanced technologies such as solid oxide fuel cells also are being actively investigated. The most promising concepts, however, are likely to be decades away from commercial reality.

Chapter 7: Status of Oxy-Combustion Capture

Introduction

In contrast to post-combustion and pre-combustion CO_2 capture technologies, which are commercial and widely used in a variety of industrial applications, oxy-combustion capture is a potential option that is still under development and not yet commercial. This chapter summarizes the current status of oxy-combustion CO_2 capture technology. The chapter again utilizes publicly available databases and project status reports maintained by the U.S. Department of Energy (DOE), the International Energy Agency's Greenhouse Gas Control Programme (IEAGHG), the Massachusetts Institute of Technology carbon sequestration program (MIT), and the Global Carbon Capture and Storage Institute (GCCSI).

In each of the sections below, the objective is to summarize not only the current status of technological developments (as of March 2010), but also the key technical barriers that must be overcome to advance oxy-combustion capture and the potential payoffs in terms of improved capture efficiency and/or reduced capture costs. Brief descriptions of new capture methods or processes not previously discussed in "Chapter 3: Overview of CO_2 Capture Technologies" also are provided.

Commercial Processes

As noted above, there are currently no oxy-combustion carbon capture systems in commercial operation. However, the critical technology of oxygen production is mature and widely used in a variety of industrial settings. The combustion of oxygen in furnaces also is practiced in industries such as glass manufacturing.

Most commercial air separation units (ASUs) employ a low-temperature cryogenic process to separate oxygen from other constituents of air (principally nitrogen and argon). The process can be scaled up or deployed in multiple trains to deliver the quantities of oxygen required for a typical coal-fired power plant. A key drawback of current ASU technology, however, is its high energy requirements, which increase with the level of oxygen purity.[77] For a typical oxyfuel plant design with 95% oxygen purity, **Table 3** earlier showed that the energy penalty for oxygen production is comparable to the penalty for amine solvent regeneration in post-combustion capture systems. Thus, for oxy-combustion carbon capture to be more economical, air separation methods are needed that are less energy-intensive than current cryogenic systems.

Full-Scale Demonstration Plants

As with post- and pre-combustion capture, to date there have been no full-scale demonstrations of oxy-combustion CO_2 capture at a power plant, nor were any oxyfuel projects selected for U.S. government support under the recent (2009) DOE Clean Coal Technology Initiatives program. A planned oxy-combustion demonstration project at a Canadian power plant also was canceled in recent years due to escalating construction costs. However, several full-scale oxy-combustion

---

[77] E. S. Rubin et al., "Estimating Future Trends in the Cost of CO_2 Capture Technologies," Report No. 2006/5, IEA Greenhouse Gas R&D Programme, Cheltenham, UK, 2006/5, January 2006.

demonstrations are still being planned outside the United States (**Table 17**), with costs to be shared between the public and private sectors. It is thus expected (but not certain) that one or more large-scale demonstrations of this technology will materialize over the next several years.

Table 17. Planned Large-Scale Demonstrations of Oxy-Combustion CO_2 Capture

Project Name and Location	Plant and Fuel Type	Expected Year of Startup	Plant Size or Capacity	Annual CO_2 Captured (10^6 tonnes)	Current Status (March 2010)
CS Energy (Australia)	Coal boiler	2011	120 MW	N/A	Under Construction
Boundary Dam (Canada)	Coal boiler	2015	100 MW	1.0	Preliminary Engineering
Vattenfall Jänschwalde (Germany)	Coal boiler	2015	500 MW	N/A	Feasibility Studies

Sources: DOE, "NETL Carbon"; IEAGHG, "CO_2 Capture"; MIT, "Carbon Capture"; GCCSI, "Strategic Analysis."

Table 17 shows that one demonstration project now under construction in Australia could begin operation as soon as next year, while two other projects would not begin operating until 2015. These oxy-combustion designs would employ a conventional ASU as the oxygen source, along with conventional flue gas cleanup systems. Potentially, a flue gas desulfurization system may be omitted to reduce costs if it is determined that sulfur oxides can be safely co-sequestered with CO_2 without compromising either the boiler or pipeline operation.

A key test for these demonstrations will be the integration of conventional ASUs to meet the oxygen needs of a large coal-fired boiler with substantial amounts of flue gas recirculation needed to control furnace temperatures. Note, too, that all of the currently planned demonstration projects are less than 200 MW in size, requiring only a single ASU train. Larger plants requiring more than 5,000 tons per day of oxygen would need multiple ASUs, adding to the complexity and cost of the oxygen delivery system.

Pilot Plant Projects

Table 18 lists one planned and two current pilot plants for testing oxy-combustion capture in an integrated system design. The two European plants currently in operation each capture over 200 tons of CO_2 per day. Vattenfall's pilot plant at the Schwarze Pumpe power station in Germany (**Figure 22**) is expected to provide critical performance data needed to design the planned demonstration plant listed earlier in **Table 17**. The oxyfuel pilot plant operated by Total in Lacq, France, is of comparable size to the Vattenfall unit. The third project in **Table 18** would supply CO_2 for a geological storage project in California as part of DOE's WESTCARB Regional Partnership. One of the two candidate sources of CO_2 is an oxy-combustion process currently under development that employs a combustor based on rocket engine technology rather than a conventional boiler system. The combustor uses a clean gaseous or liquid fuel to produce steam and CO_2 at high pressure.[78] A final decision on the availability and use of this process for the WESTCARB project is still pending as of this writing.

---

[78] Clean Energy Systems, "Zero-Emission Power Plants (ZEPP)," Clean Energy Systems, (continued...)

Table 18. Pilot Plant Projects with Oxy-Combustion CO_2 Capture

Project Name and Location	Plant and Fuel Type	Expected Year of Startup	Plant Size or Capacity	Annual CO_2 Captured (10^6 tonnes)	Current Status (March 2010)
Schwarze Pumpe (Germany)	Coal Boiler	2008	30 MW_{th} (~10 MW)	0.075	Operational
Total Lacq (France)	Natural Gas Boiler	2009	30 MW_{th} (~10 MW)	0.075	Operational
WESTCARB[a] (California)	Gaseous or liquid fueled combustor[a]	2012	~50 MW	0.25	Preliminary Engineering

Sources: DOE, "NETL Carbon"; IEAGHG, "CO_2 Capture"; MIT, "Carbon Capture"; GCCSI, "Strategic Analysis."

Note: MW = net electrical megawatts output; MW_{th} = gross thermal energy input (megawatts).

a. One of two proposed CO_2 sources for this project; final decision expected mid-2010.

Figure 22. Oxy-Combustion Pilot Plant Capturing CO_2 from the Flue Gas of a Coal-Fired Boiler at the Schwarze Pumpe Power Station in Germany

Source: Photo courtesy of Vattenfall.

Not included in **Table 18** are other pilot-scale facilities around the world that are also used to test various components of an oxy-combustion system, such as the 30 MW_{th} Clean Energy Development Facility of Babcock and Wilcox.[79] Similarly, Air Products is operating a pilot plant in Maryland that uses an experimental ion transport membrane (ITM) system for oxygen

---

(...continued)

http://www.cleanenergysystems.com/technology html.

[79] K. J. McCauley et al., "Commercialization of Oxy-Coal Combustion: Applying the Results of a Large 30MWth Pilot Project," Proc. *9th International Conference on Greenhouse Gas Control Technologies, Nov 16-20, 2008*, Washington, DC. Note: MW_{th} = gross thermal energy input (megawatts).

production, rather than a conventional cryogenic ASU.[80] That system, depicted in **Figure 23**, is one of several new technologies under development that promise to deliver lower-cost oxygen.

While not a CO_2 capture technology, oxygen production is nonetheless the major cost and energy penalty item of an oxy-combustion system. For that reason, advanced methods of oxygen production are discussed in the following section on laboratory- and bench-scale developments.

Figure 23. The Ion Transport Membrane (ITM) Oxygen Production Technology Being Developed by Air Products

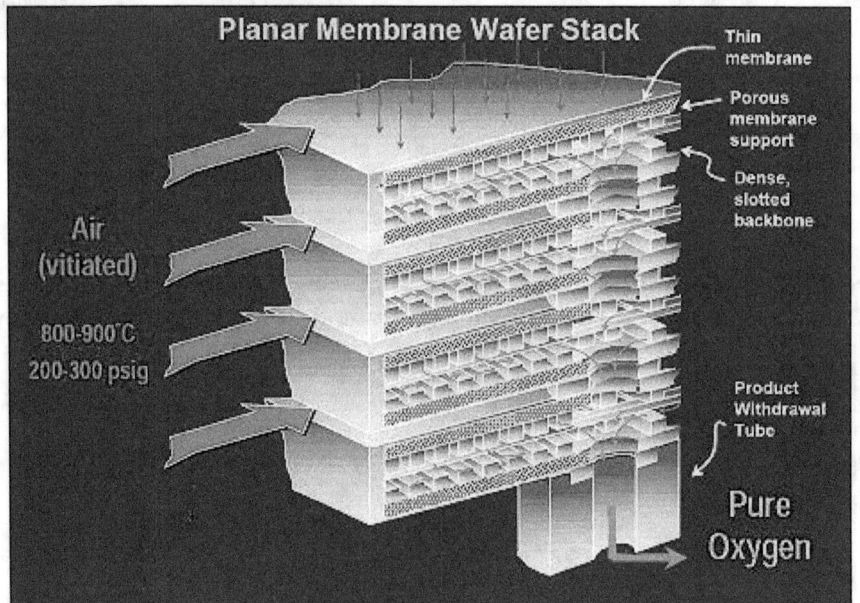

Source: P. A. Armstrong et al., "ITM Oxygen for Gasification," Gasification Technologies Council, Gasification Technologies Conference, Washington, DC, October 3-6, 2004.

Laboratory- or Bench-Scale Developments

Laboratory- and bench-scale R&D related to oxy-combustion is found worldwide and is currently focused mainly in the following areas:

- understanding oxy-combustion burner and boiler characteristics and their interactions with the overall system;

- design of innovative oxy-combustion burners for new and retrofit applications;

- development of improved flue gas purification technologies for oxy-fired systems;

- development of lower-cost, high-efficiency oxygen production units; and

- development of novel concepts such as chemical looping combustion.

---

[80] Process Worldwide, "Rising Demand is Shaking up the Staid Business of Air Separation," http://www.process-worldwide.com/.

Topics of study include investigations into the fundamental mechanisms that affect the performance and design of oxygen-fired boiler systems, such as studies of oxy-combustion flame characteristics, burner design, and fuel injection systems. Because of the high temperatures associated with oxygen combustion, the development of advanced boiler materials is another focus of research. In a number of areas, small-scale experiments are coupled with computational fluid dynamic (CFD) modeling studies of oxy-combustion processes.

The development of advanced flue gas purification systems also is being pursued to find better, lower-cost ways to remove contaminants such as sulfur oxides, nitrogen oxides, and trace elements such mercury. The ability to remove such pollutants during the CO_2 compression process is one of the promising recently reported innovations being studied.[81]

There is a large body of technical literature that discusses and documents in detail the range of laboratory- and bench-scale R&D activities and challenges in oxy-combustion CO_2 capture.[82] The remainder of this section elaborates briefly on the two areas believed to offer the greatest promise for lower-cost capture.

Advanced Oxygen Production Methods

Current commercial technology uses low-temperature (cryogenic) separation methods to produce high-purity oxygen. An alternative that promises a lower energy penalty and lower cost is the ion transport membranes (ITM) system mentioned earlier. Here, thin nonporous membranes are used to separate oxygen from air at high temperature and pressure, as seen in **Figure 23**. As with other membrane-based systems, the separation works on the principle of an oxygen pressure difference on either side of the membrane. The higher the pressure difference, the better the separation. The goal of R&D at Air Products, Inc., is to produce ITM oxygen at one-third the cost and energy requirement of current cryogenic ASUs.[83] IGCC systems and other gasification-based processes are currently the most attractive applications for ITM oxygen, since these processes already operate at the high pressures required by ITM technology. Oxy-combustion applications, however, would require the development of pressurized combustion systems in order to take full advantage of ITM oxygen production.

Unlike ITMs, which separate oxygen based on a pressure differential, the oxygen transport membrane (OTM) concept utilizes the chemical potential of oxygen as the driving force. The potential advantage of this approach is that it can be integrated directly into a boiler, with air on one side of the membrane and fuel combustion on the other side. Combustion decreases the oxygen concentration, which increases the chemical potential difference that drives O_2 through the membrane. This process is still in the early stages of development.[84]

---

[81] K. White et al., "Purification of Oxyfuel-Derived CO_2," *International Journal of Greenhouse Gas Control*, vol. 4, no. 2 (2010), pp. 137-142.

[82] R. Allam, "Carbon Dioxide Capture Using Oxyfuel Systems," Carbon Capture-Beyond 2020, U.S. Department of Energy, Office of Basic Energy Research, March 4-5, 2010, Gaithersburg, MD; International Energy Agency Greenhouse Gas R&D Programme, "Oxy-fuel Combustion Network," at http://www.co2captureandstorage.info/networks/oxyfuel htm.

[83] P. A. Armstrong et al., "ITM Oxygen."

[84] DOE, "Carbon Dioxide Capture."

Other new oxygen production methods being investigated use solid sorbents to absorb O_2 from air. The sorbent material is then transferred to another vessel where it is heated, releasing the O_2. This is fundamentally the same approach discussed in "Chapter 5: Status of Post-Combustion Capture" and "Chapter 6: Status of Pre-Combustion Capture" for CO_2 capture using solid sorbents. For O_2 production the sorbent material and process conditions are different. The process called ceramic autothermal recovery uses the mineral perovskite. It releases heat while adsorbing O_2 from air, which potentially could be used along with heat from power plant flue gases to reduce the overall energy penalty of oxygen production. Another sorbent being investigated is manganese oxide, which absorbs O_2 from high pressure air passed over the sorbent. This technology is potentially easier to build and lower in cost.[85] Until a larger-scale process is developed and tested, however, cost estimates remain highly uncertain.

Chemical Looping Combustion

Another novel oxy-combustion technology being developed is called chemical looping combustion (CLC). This is similar to the sorbent-based O_2 production method discussed above. Here, however, the O_2-carrying sorbent—typically a metal oxide—is contacted with a fuel, so that combustion occurs rather than a simple release of oxygen. The resulting exhaust stream contains only carbon dioxide and water vapor, as in other oxy-combustion schemes. A schematic of this concept is shown in **Figure 24**.

Figure 24. Schematic of a Chemical Looping Combustion System

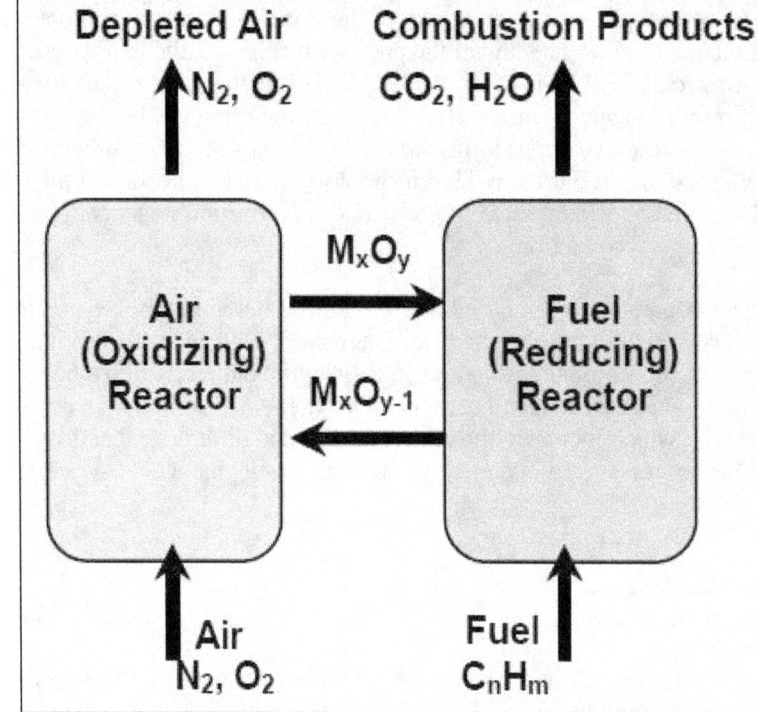

Source: DOE, "Carbon Dioxide Capture."

---

[85] DOE, "Carbon Dioxide Capture."

Chemical looping has the potential to make carbon capture significantly cheaper than current systems, but is still at an early stage of development, with challenges in materials handling and oxygen carrier selection that have not yet been solved. Currently the largest chemical looping combustor is a 120-kilowatt unit being tested in Austria.[86] Projects funded by the U.S. Department of Energy include two chemical looping tests, one by Alstom using calcium compounds as an oxygen carrier, the other by Ohio State University using an iron oxide carrier. Alstom currently has a 65 kW test reactor and plans to have a 3 MW pilot plant online in late 2010.[87]

Conceptual Design Stage

As with pre-combustion CO_2 capture systems, a substantial amount of current activity on oxy-combustion capture is still at the conceptual design stage, positing and analyzing alternative system configurations that maximize overall efficiency and minimize estimated cost. Conceptual designs encompass a broad range of fuels and power systems. Many of these designs include advanced component technologies and heat integration schemes that do not currently exist, but which illustrate the potential for process improvements.

For example, a proposed novel oxy-combustion cycle for natural gas-fired power plants combines an oxygen transport membrane with advanced heat integration in a reactor design (**Figure 25**) that theoretically achieves 85% to 100% CO_2 capture with a plant efficiency much higher than a current NGCC plant with CO_2 capture.[88] Other oxy-combustion designs for combined cycle power plants utilize CO_2 instead of air to generate power from advanced gas turbines, or employ ITM technology to achieve high-efficiency power generation with high CO_2 capture.[89] All of these advanced concepts, however, require the (costly) development and integration of advanced technologies that do not yet exist and which may have only limited market potential. Thus, despite their theoretical advantages, it appears unlikely that such concepts will advance to the later stages of technology development any time soon.

Other conceptual designs for coal-fired power plants[90] seek improved methods of heat and process integration to improve overall plant efficiency using conventional technology for power generation and oxygen production. More advanced concepts envision pressured combustion with oxygen as a preferred approach for achieving high efficiency along with lower-cost CO_2 capture. These analyses based on thermodynamics and optimization methods are useful for identifying the most promising concepts to consider for further development.

---

[86] P. Kolbitsch et al., "Operating Experience with Chemical Looping Combustion in a 120 kW Dual Circulating Fluidized Bed (DCFB) Unit," *Energy Procedia*, vol. 1 (2009), pp. 1465 -1472.

[87] H. E. Andrus et al., "Alstom's Calcium Oxide Chemical Looping Combustion Coal Power Technology Development" Proc. *34th International Technical Conference on Clean Coal & Fuel Systems, May 31-June 4, 2009*, Clearwater, FL.

[88] R. Anantharaman, O. Bolland, and K. I. Asen, "Novel Cycles for Power Generation with CO2 Capture using OMCM Technology," *Energy Procedia*, vol. 1, no. 1 (2009), pp. 335-342.

[89] Metz, "Special Report." O. Bolland, "Outlook for Advanced Capture Technology," presented at *The 9th International Conference on Greenhouse Gas Control Technologies, November, 16-20, 2008*, Washington, DC.

[90] K. E. Zanganeh and A. Shafeen, "A Novel Process Integration, Optimization and Design Approach for Large-Scale Implementation of Oxy-Fired Coal Power Plants with CO_2 Capture," *International Journal of Greenhouse Gas Control*, vol. 1 (2007), pp. 47-54l; Allam, "Oxyfuel Systems."

Figure 25. A Proposed Oxygen-Mixed Conduction Membrane Reactor Design for a Natural Gas-Fired Power Plant

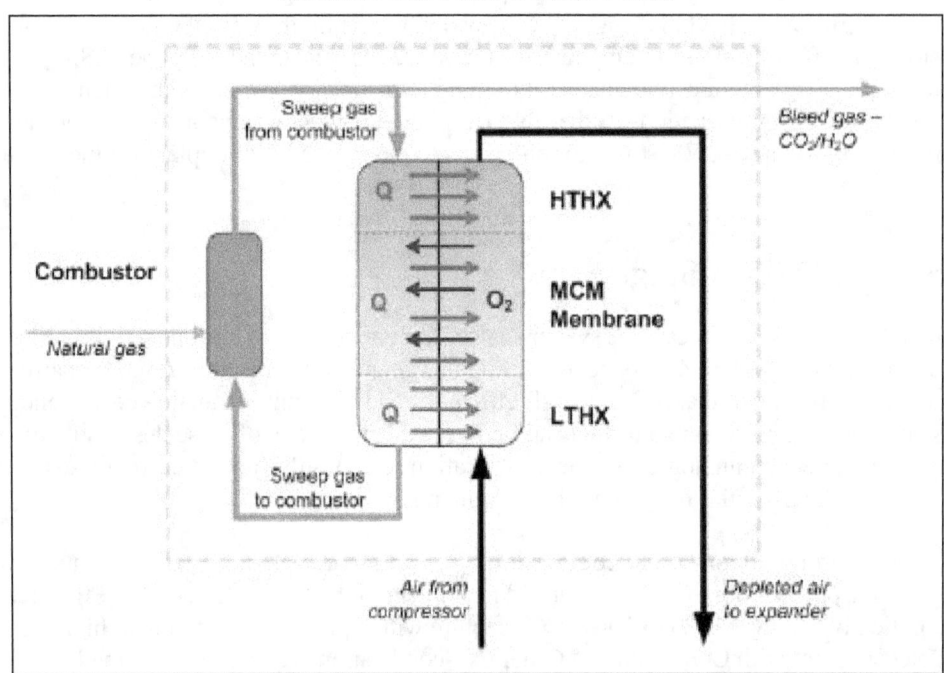

Source: R. Anantharaman, O. Bolland, and K. I. Asen, "Novel Cycles for Power Generation with CO_2 Capture using OMCM Technology," *Energy Procedia*, vol. 1, no. 1 (2009), pp. 335-342.

Conclusion

This chapter has reviewed and summarized a range of R&D activities underway to develop oxy-combustion CO_2 capture as an alternative to post-combustion capture, especially for coal-fired boilers. Some of these activities are similar in nature to those for post-combustion and pre-combustion capture insofar as they involve the same basic concepts, such as the use of membrane separation processes. In the context of oxy-combustion systems, however, the most compelling need—and a major focus of R&D—is for improved, lower-cost processes to deliver large quantities of high-purity oxygen, the major cost item in current oxyfuel schemes. To the extent that oxy-combustion systems are able to transport and sequester multi-pollutant gas streams, including pollutants like SO_2 and NO_x, costs can be further reduced by avoiding the need for additional gas cleaning equipment to remove such pollutants. At the conceptual level, advanced plant designs employing new plant integration concepts and advanced technologies such as chemical looping combustion are also being actively investigated and many appear promising. Because they are at the earliest stages of development, however, it remains to be seen which if any of these concepts eventually develops into a viable commercial technology.

Chapter 8: Cost and Deployment Outlook for Advanced Capture Systems

Introduction

This chapter addresses two key questions not addressed in the previous chapters: (1) How much cost reduction and performance improvement is expected from the CO_2 capture technologies now under development? and (2) When will these technologies be available for commercial use? To address the first question, this chapter shows results from recent studies by DOE and others of projected cost reductions for power plants with advanced capture systems. To address the second question, this chapter presents a set of technology roadmaps and deployment scenarios developed by governmental and private organizations involved in CO_2 capture technology R&D. "Chapter 9: Lessons from Past Experience" reviews past experience in other R&D programs to develop advanced capture technologies for power plant emissions.

Projected Cost Reductions for CO_2 Capture

Table 4 earlier summarized the range of cost estimates for power plants using current technology for CO_2 capture and storage. Other sources discuss in detail the many factors that affect such estimates.[91] In the context of the present report, it is especially important to emphasize the uncertainty inherent in any cost estimate for a technology that has not yet been built, operated, and replicated at a commercial scale. In general, the farther away a technology is from commercial reality, the cheaper it tends to look. This is illustrated graphically in **Figure 26**, which depicts the typical trend in cost estimates for a technology as it advances from concept to commercial deployment.

Keeping in mind this uncertainty, this section summarizes the results of several recent studies that estimated potential cost reductions from technology innovations both in CO_2 capture processes and in other power plant components that influence CO_2 capture cost. These studies employ two conceptually different methods of estimating future costs. The "bottom up" method uses engineering analysis and costing to estimate the total cost of a specified advanced power plant design. In contrast, the "top down" method uses learning curves derived from past experience with similar technologies to estimate the future cost of a new technology based on its projected installed capacity at some future time. The latter parameter represents the combined effect of all factors that influence historically observed cost reductions (including R&D expenditures, learning-by-doing, and learning-by-using).

---

[91] Metz, "Special Report." E. S. Rubin et al., *The Effect of Government Actions on Environmental Technology Innovation: Applications to the Integrated Assessment of Carbon Sequestration Technologies*, report from Carnegie Mellon University, Pittsburgh, PA, to U.S. Department of Energy, Germantown, MD, p. 153, January 2004.

Figure 26. Typical Trend in Cost Estimates for a New Technology as It Develops from a Research Concept to Commercial Maturity

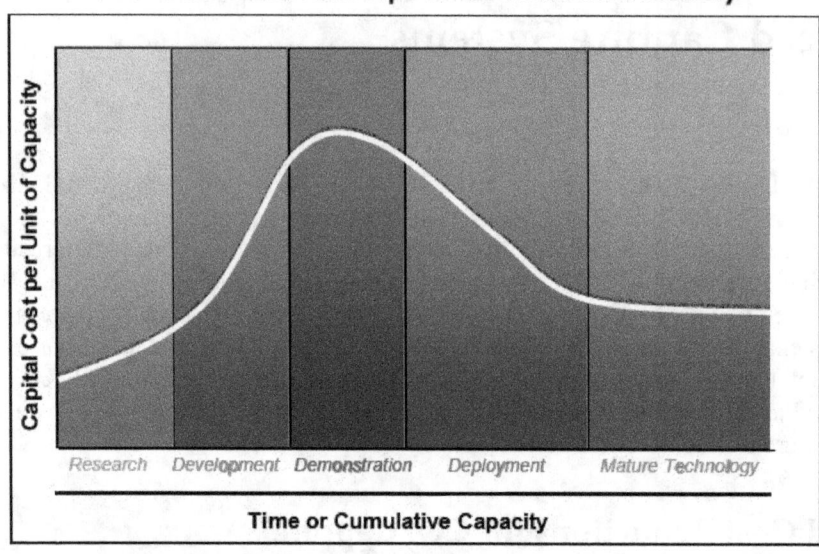

Source: Adapted from S. Dalton, "CO₂ Capture at Coal Fired Power Plants—Status and Outlook," The 9th International Conference on Greenhouse Gas Control Technologies, Washington, DC, November, 16-20, 2008.

Results from Engineering-Economic Analyses

Figure 27 shows the results of a 2006 analysis by DOE of potential advances in the major CO_2 capture routes. Results are shown for pulverized coal (PC) plants and integrated gasification combined cycle (IGCC) plants. The bars in **Figure 27** show the percent increase in the total cost of electricity (COE) compared to the same plant type without CO_2 capture. As more advanced technologies are implemented, the incremental cost is reduced significantly. On an absolute basis, the total cost of electricity generation falls by 19% for the IGCC cases and by 28% for the PC cases. The biggest cost reductions come in the final steps for each plant type. However, the technologies in those cases are still in the early stages of development, including advanced solid sorbents for CO_2 capture membrane systems for water-gas shift reactors and chemical looping for oxygen transport. As suggested earlier in **Figure 26**, cost estimates for these cases are the least reliable and most likely to escalate as the technology approaches commercialization.

The 2006 DOE analysis also included four oxy-combustion cases (not shown in **Figure 27**) in which the COE for an advanced system fell by 19% (from a 50% increase in COE for a current supercritical PC plant, to a 21% increase for advanced SCPC with ITM oxygen production). Because oxy-combustion systems are still under development and not yet demonstrated at a commercial scale, assumed plant configurations and cost estimates for these systems are more uncertain and variable than for current pre- and post-combustion systems. For example, while some studies show oxy-combustion for new power plants to be somewhat lower in cost than post-combustion capture,[92] others report it to be higher in cost.[93] There is general agreement, however, that continued R&D can reduce the future cost of these systems.

---

[92] U.S. Department of Energy, *Pulverized Coal Oxycombustion Power Plants: Volume 1, Bituminous Coal to Electricity*, Report No. DOE/NETL-2007/1291, National Energy Technology Laboratory, Pittsburgh, PA, August 2008.

[93] Metz, 'Special Report.'

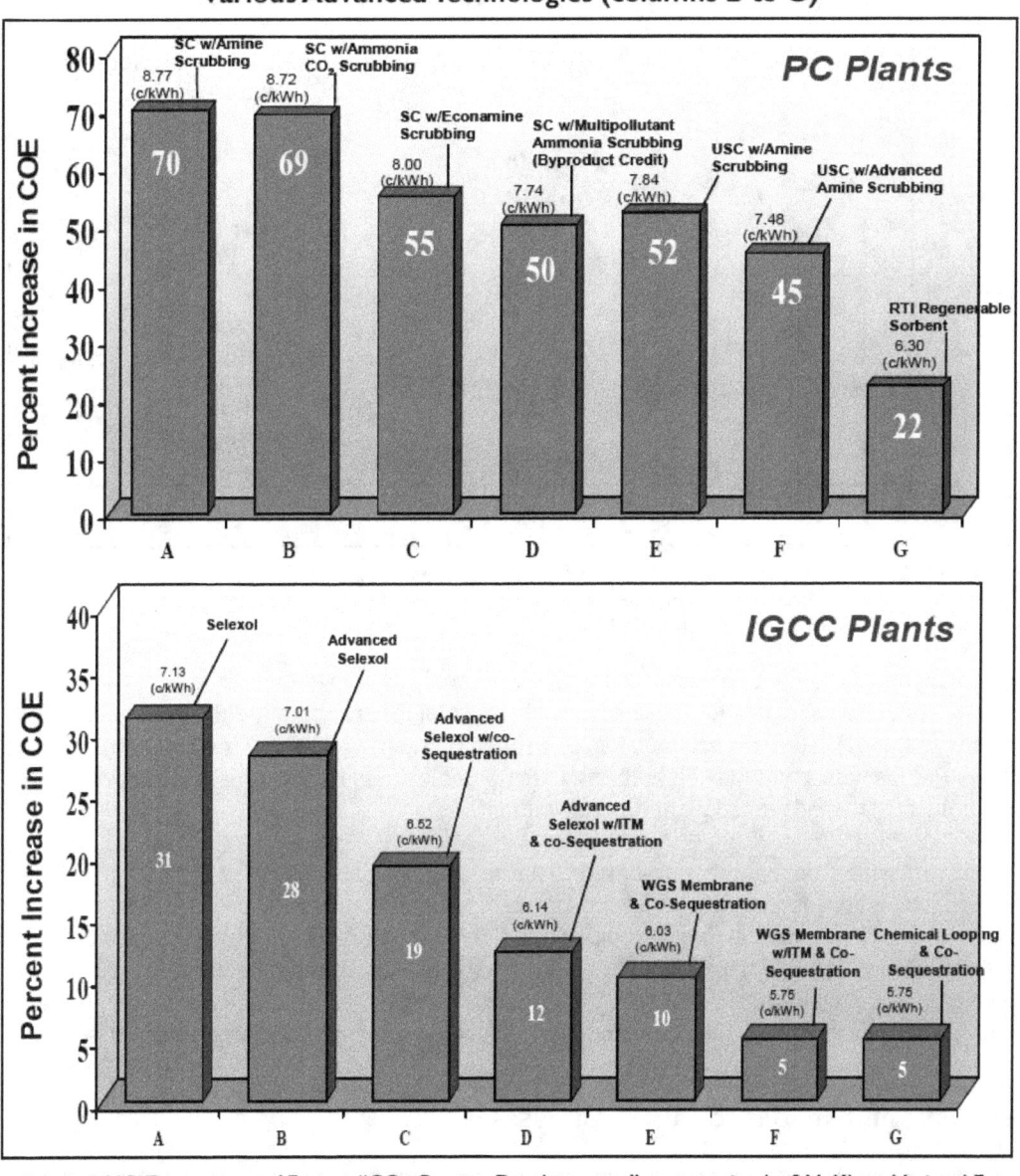

Figure 27. Cost of Electricity (COE) Increases for Power Plants with CO₂ Capture and Storage Using Current Technology (column A) and Various Advanced Technologies (columns B to G)

Source: U.S. Department of Energy, "CO₂ Capture Developments," presentation by S.M. Klara, National Energy Technology Center and Office of Fossil Energy, Strategic Initiatives for Coal, Queenstown, MD, December 2006.

Notes: The value of total COE appears at the top of each column. Abbreviations: SC = supercritical; USC = ultrasupercritical; RTI = Research Triangle Institute; ITM = ion transport membrane; WGS = water gas shift.

Figure 28 shows a more recent (2010) DOE analysis of potential reductions in capture cost from sustained R&D. Here, the total cost of a new supercritical PC plant with CCS declines by 27% while the IGCC plant cost falls by 31%. Thus, the future IGCC plant with CCS costs 7% less than the current plant without capture. For the PC plant the CCS cost penalty falls by about half in this analysis.

Figure 28. Current Cost of Electricity (COE) for IGCC and PC Power Plants with and without CO₂ Capture and Storage (CCS), Plus Future Costs with Advanced Technologies from R&D

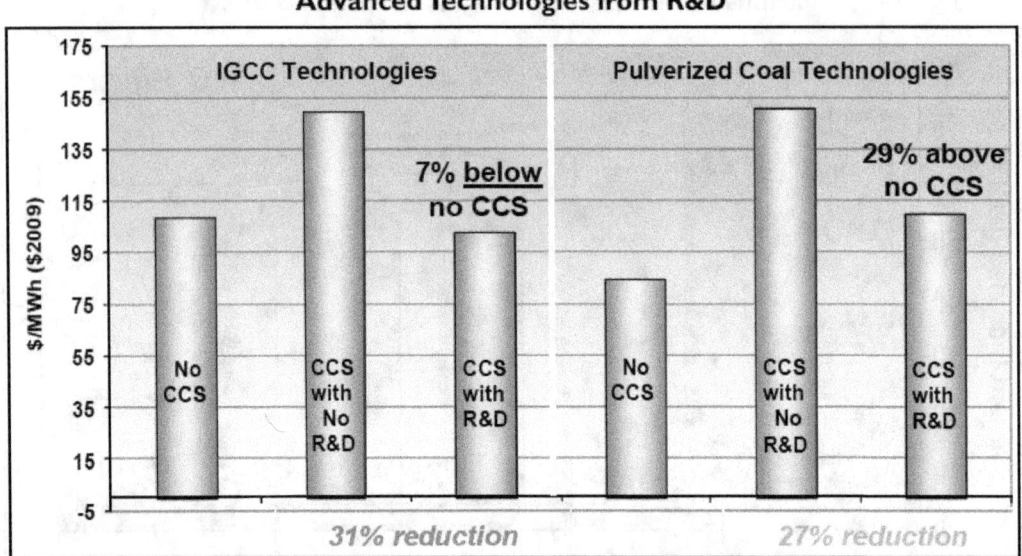

Source: U.S. Department of Energy, "Carbon Dioxide Capture and Storage (CCS)," CCS Briefing to Senate Energy and Natural Resources Committee, by S. M. Klara, National Energy Technology Center and Office of Fossil Energy, Washington, DC, March 5, 2010.

Since many of the components assumed in the DOE analysis are still at early stages of development, cost estimates for these advanced technologies are again highly uncertain. Nonetheless, these estimates can be taken as a rough (perhaps optimistic) indication of the potential cost savings that may be realized. Other organizations have estimated similar cost reductions for other advanced plant designs with CCS.[94]

Typically missing from engineering-based cost estimates such as these is an indication of the time frame in which advanced technologies are expected to be in commercial use. This is especially problematic for environmental technologies like CO₂ capture processes, since the market for such systems depends mainly on government policies that require or incentivize their use. An alternative approach to forecasting technology costs, based on learning or experience curves, comes closer to providing a temporal dimension together with cost estimates, as discussed below.

Results from Experience Curve Analyses

As noted earlier, the "top down" approach to cost estimation models the future cost of power plants with CCS as a function of the total installed capacity of such plants. While time is not an explicit variable, it is implied by the choice of CCS plant capacity that is projected. The future cost reductions shown in **Figure 29** are from a detailed analysis that applied historical learning rates for selected technologies to the components of four types of power plants with CO₂ capture (PC, NGCC, IGCC, and oxyfuel).[95] The component costs were then summed to estimate the

---

[94] E.g., Metz, "Special Report."

[95] E. S. Rubin et al., "Use of Experience Curves to Estimate the Future Cost of Power Plants with CO2 Capture," *International Journal of Greenhouse Gas Control*, vol. 1, no. 2 (2007), pp. 188-197.

future cost of the overall power plant as a function of new plant capacity. The analysis also considered uncertainties in key parameters, including potential increases in cost during early commercialization.

Figure 29. Projected Cost Reductions for Four Types of Power Plants with CO_2 Capture Based on Experience Curves for Major Plant Components

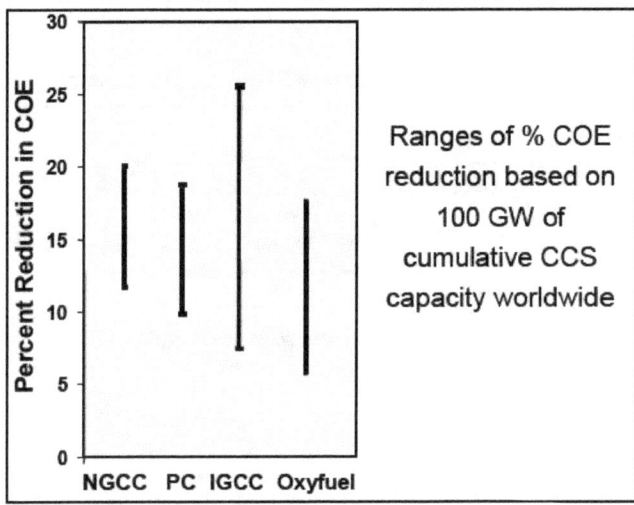

Ranges of % COE reduction based on 100 GW of cumulative CCS capacity worldwide

Source: E. S. Rubin et al., "Use of Experience Curves to Estimate the Future Cost of Power Plants with CO_2 Capture," *International Journal of Greenhouse Gas Control*, vol. 1, no. 2 (2007), pp. 188-197.

Figure 29 shows the resulting ranges of cost reduction estimated for each of the four types of power plants with CO_2 capture after an assumed deployment of 100,000 MW for each system worldwide (100,000 MW was the total installed capacity of flue gas desulfurization systems approximately 20 years after that technology was first introduced at U.S. power plants). Note that these results reflect the maturity of each plant type as well as the CO_2 capture system. Thus, the IGCC plant—whose principal cost components are less mature than those of combustion-based plants—shows the largest potential for overall cost reductions. The combustion-based plants show a smaller potential, since most of their components are already mature and widely deployed. In all cases, however, the incremental cost of CO_2 capture system falls more rapidly than the cost of the overall plant.

Note that the high end of the cost reduction ranges in **Figure 29** is similar to DOE's "bottom up" estimates shown in **Figure 27**. The low end of the ranges, however, is smaller by factors of two to three. That result suggests a more gradual rate of cost reductions from continual improvements to capture technologies as CCS is more widely deployed.

Roadmaps for Capture Technology Commercialization

This section looks at estimated timetables for the development and commercialization of CO_2 capture systems. Such "roadmaps" have been developed by a number of governmental and private organizations involved in CO_2 capture technology R&D. They provide a useful perspective on the time frame in which improved or lower-cost capture systems are expected to become commercial and available for use at power plants and other industrial facilities.

The DOE Roadmap

As part of its Carbon Sequestration Program, the U.S. Department of Energy (DOE) has developed and periodically updates a roadmap displaying the projected timetable for major program elements, including CO_2 capture technology development. **Figure 30** shows an excerpt from the most recent DOE roadmap published in 2007. **Figure 31** shows a more detailed timeline for advanced CO_2 capture technologies applied to existing plants.

Figure 30. The DOE Carbon Sequestration Program Roadmap from 2012 to 2022

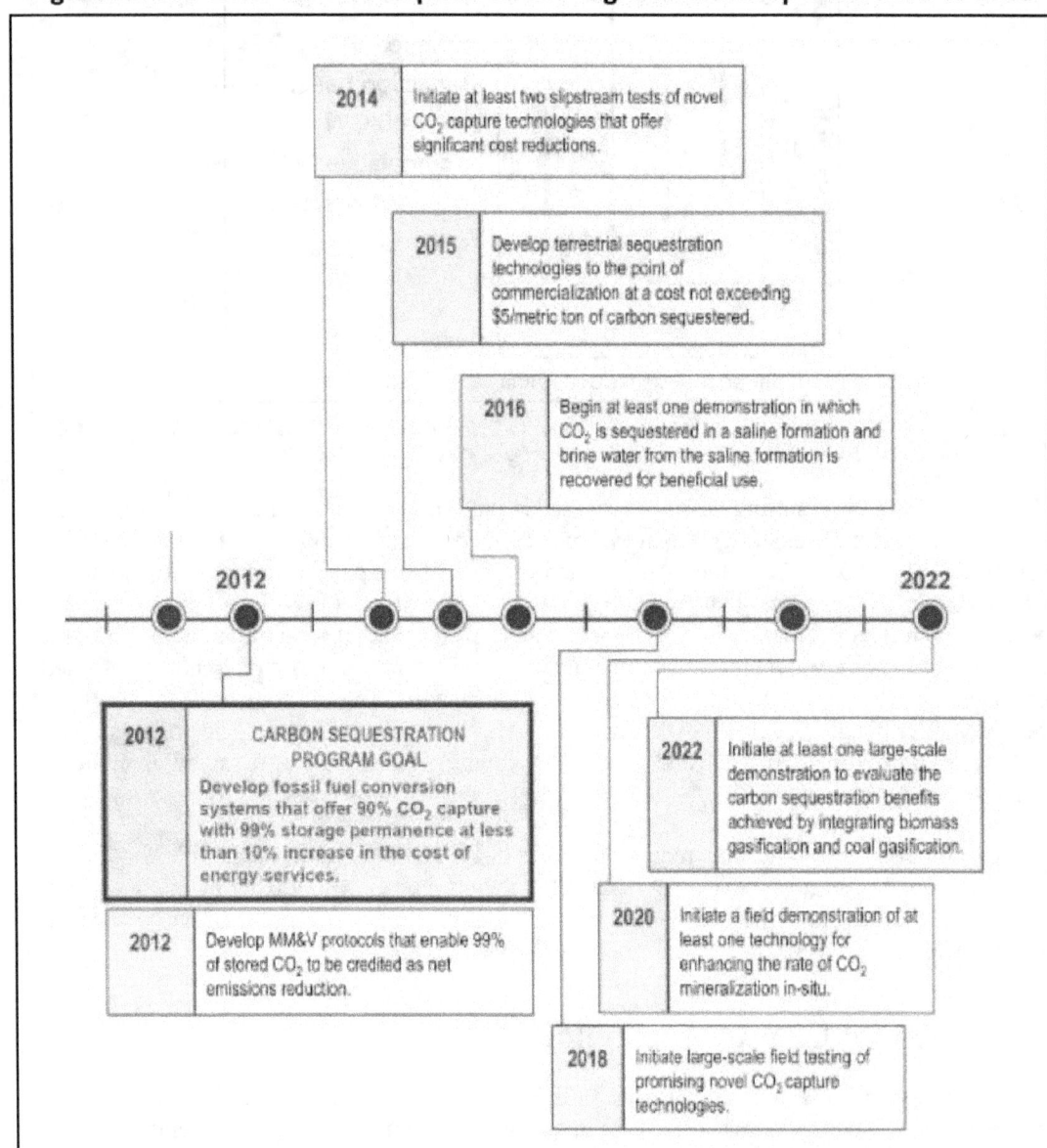

Source: U.S. Department of Energy, *Carbon Sequestration Technology Roadmap and Program Plan*, National Energy Technology Laboratory, Pittsburgh, PA, 2007.

Figure 31. DOE's Timeline from R&D to Commercial Deployment of Advanced Post-Combustion Capture Technologies for Existing Power Plants

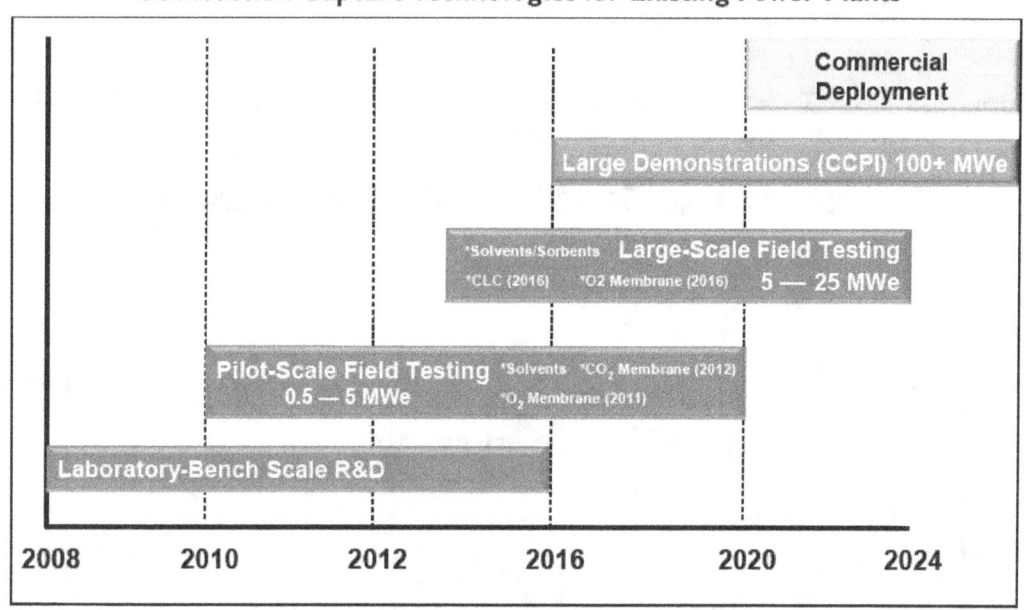

Source: Ciferno, "DOE/NETLs Existing Plants."

The 2007 DOE roadmap has milestones extending to 2022. The more recent roadmap for advanced post-combustion capture systems in **Figure 31** extends beyond 2024. It anticipates commercial deployment of advanced technologies in 2020, with full-scale demonstrations beginning four years earlier, in 2016.[96] Laboratory- and bench-scale R&D would, on average, advance to pilot-scale testing after about two years, with subsequent pilot plant testing and scale-up prior to large-scale demonstrations.

The Electric Power Research Institute (EPRI) carries out R&D on behalf of member utility companies. EPRI-supported projects include development and testing of advanced carbon capture technologies. **Figure 32** shows a roadmap developed jointly between EPRI and the Coal Utilization Research Council (CURC), an industry advocacy group that promotes the efficient and environmentally sound use of coal. Recent updates to this roadmap call for four demonstrations of IGCC with CCS by 2025, including the FutureGen project (noted earlier in "Chapter 6: Status of Pre-Combustion Capture"), plus nine demonstrations of combustion with CCS by 2025.[97] Like the DOE plan, the CURC-EPRI roadmap expects CO_2 capture systems for power plants to be commercial by 2020. That roadmap, however, shows a heavier reliance on continued improvements to technologies that are already at the advanced stages of development.

EPRI researchers also have put forth a timeline for carbon capture developments based on the Technology Readiness Levels (TRLs) described earlier in "Chapter 4: Stages of Technology Development." This timeline, shown in **Figure 33**, characterizes most systems being developed today at TRLs 5 through 7. It shows activity at TRL 8 (equivalent to large-scale demonstration projects) beginning in 2010, with commercial-scale plants (TRL 9) coming online by 2018. This

---

[96] J. P. Ciferno, "DOE/NETLs Existing Plants CO_2 Capture R&D Program," Proc. *Carbon Capture 2020 Workshop, October 5-6, 2009*, College Park, MD.

[97] Coal Utilization Research Council (CURC), Clean Coal Technology Roadmap, Washington, DC (2009).

implies a 10- to 15-year development schedule from concept to commercialization. EPRI acknowledges, however, that this schedule represents an aggressive and well-funded program of research, development, and deployment.

Figure 32. Steps in Technology Validation and Scale-Up Projects to Meet CURC-EPRI Roadmap Goals for Advanced Coal Technologies with CCS

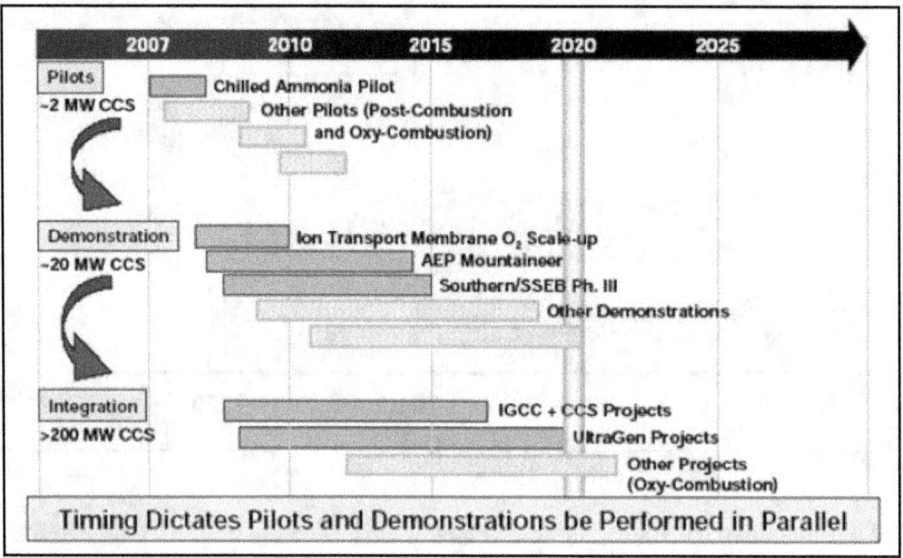

Source: CURC, "Clean Coal."

Figure 33. EPRI Projections of Capture Technology Development Based on Technology Readiness Levels (TRLs)

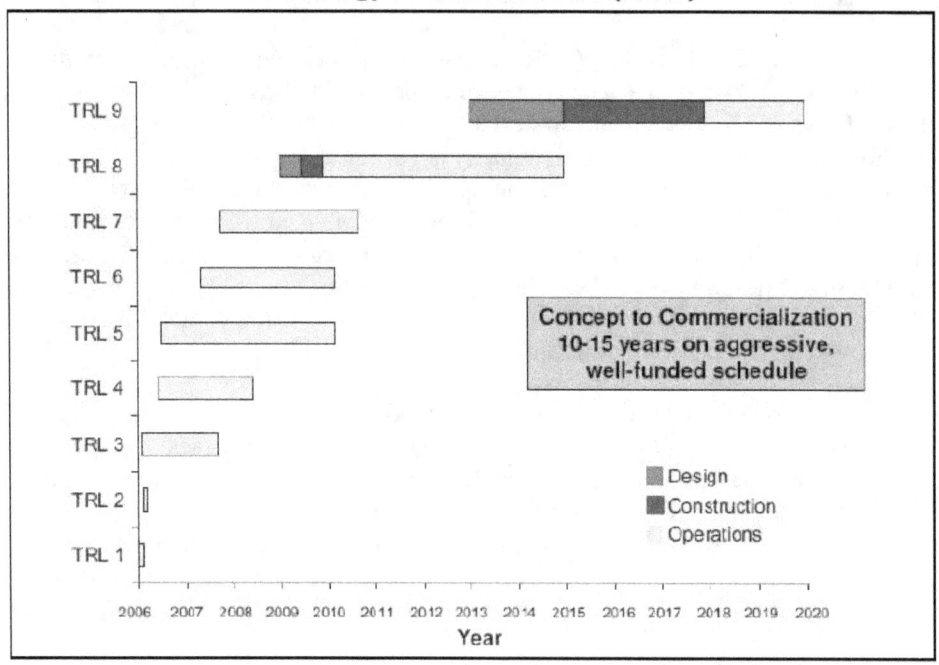

Source: Bhown and Freeman, "Assessment of Post-Combustion."

The CSLF Roadmap

The Carbon Sequestration Leadership Forum (CSLF) is an international climate change initiative (at the ministerial level) focused on the development of improved cost-effective technologies for CO_2 capture and storage. Its mission is to facilitate the development and deployment of such technologies via collaborative efforts.

The CSLF roadmap in **Figure 34** sets out development goals in three time periods: 2009-2013, 2014-2020, and 2020 and beyond. For CO_2 capture, the goal for the first stage is "development of low-cost and scalable carbon capture technologies." Goals for the second stage involve full-scale demonstrations of these technologies, while the goal for 2020 and beyond is to have these technologies deployed commercially.[98] The roadmap also lays out goals for CO_2 transport and storage and for the development of integrated full-scale CCS projects by 2013. As an international organization, the CSLF does not itself provide funding for CO_2 capture R&D, but rather relies on country-level support for such projects.

Figure 34. Key Milestones in the CSLF Technology Roadmap

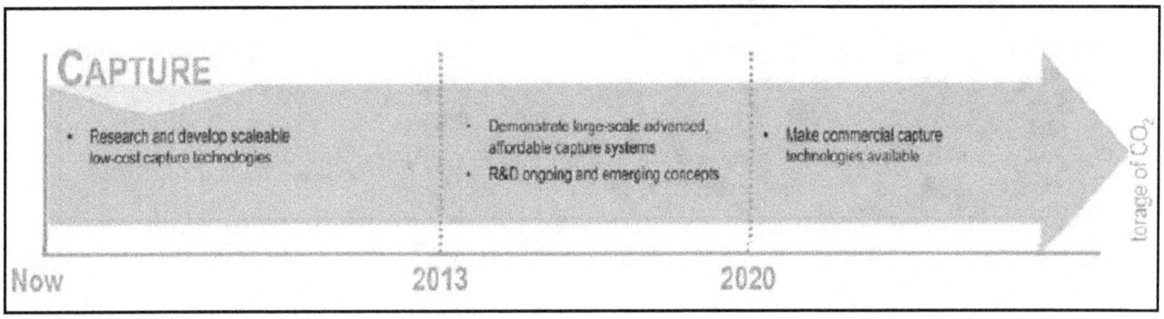

Source: Carbon Sequestration Leadership Forum, "Carbon Sequestration Leadership Forum Technology Roadmap: A Global Response to the Challenge of Climate Change," http://www.cslforum.org/publications/documents/CSLF_Techology_Roadmap.pdf.

Other Roadmaps and Milestones

Several other international groups and organizations have set goals and targets for the demonstration, commercialization, and deployment of CO_2 capture and storage systems. At its 2008 summit meeting in Japan, the Group of Eight (G8)—representing the governments of Canada, France, Germany, Italy, Japan, Russia, the United Kingdom, and the United States—committed to "strongly support the launching of 20 large-scale CCS demonstration projects globally by 2010, ... with a view to beginning broad deployment of CCS by 2020."[99] This action was based on recommendations of the CSLF and the International Energy Agency (IEA).

In conjunction with its global energy modeling activities, IEA also has published a CCS roadmap calling for increasing numbers of pilot and demonstration plants worldwide through 2035.[100] To

---

[98] Carbon Sequestration Leadership Forum, "Technology Roadmap."

[99] Group of Eight 2008, G8 Summits Hokkaido Official Documents—Environment and Climate, http://www.g7.utoronto.ca/summit/2008hokkaido/2008-climate.html.

[100] International Energy Agency, "Technology Roadmap: Carbon Capture and Storage," http://www.iea.org/papers/2009/CCS_Roadmap.pdf.

support the commercialization of CCS globally, the IEA sees a requirement for about 30 such new-build pilot and demonstration projects in the 2020-2025 time frame, an additional 100 projects in 2025-2030, and about 40 more in 2030-2035. A majority of early large-scale projects would take place in OECD countries, but after 2030 non-OECD countries would take the lead in commercializing CCS plants.

A number of countries also have developed national plans or projections for CCS. **Figure 35** shows the R&D needs and timetable for CO_2 capture systems identified in a CCS roadmap for Canada.

Figure 35. Capture System R&D Needs in the CCS Roadmap for Canada

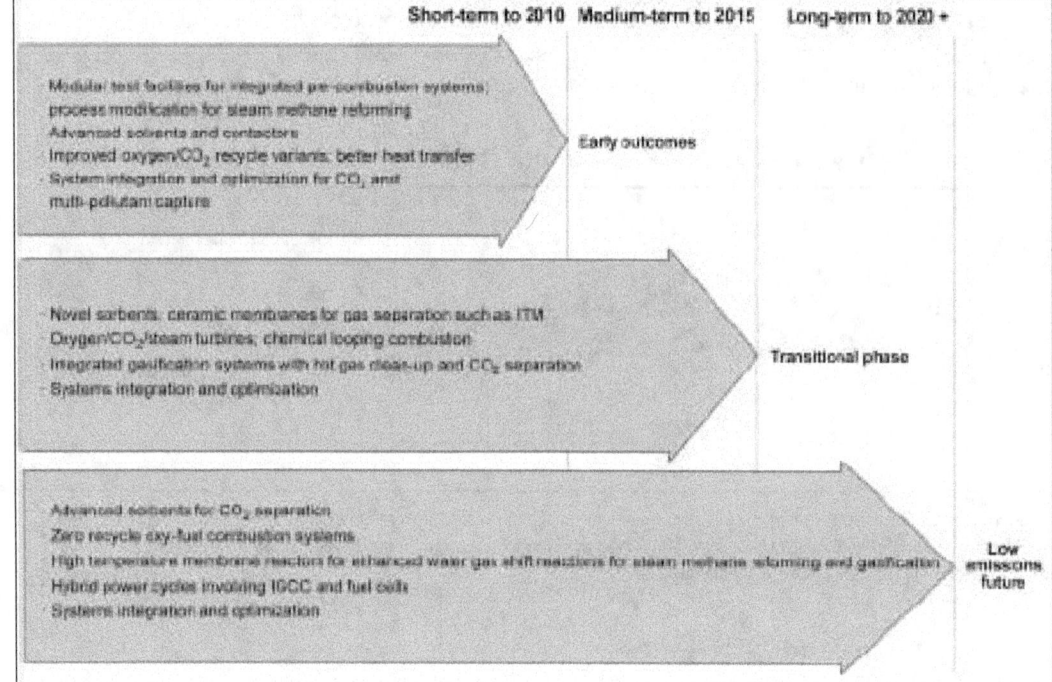

Source: Natural Resources Canada, Canada's Carbon Dioxide Capture and Storage Technology Roadmap, March 2006.

Scenarios for CCS Deployment

Recent studies include a wide range of scenarios modeled by different groups to predict the consequences of national and international policies to mitigate global climate change.[101] These studies typically assume that CCS is available for deployment at power plants and other industrial facilities by at least 2020. While the future cost of CCS assumed in different models is not readily available, the scenario results indicate widely differing projections of CCS deployment.

For example, **Figure 36** shows results from five different models used to project the U.S. energy mix in 2050 in response to policy scenarios requiring national reductions in greenhouse gas

---

[101] National Research Council, "America's Climate Choices." Metz, "Climate Change 2007."

emission of 50% to 80% below 1990 levels.[102] Results for the five models for the year 2035 show deployment of CCS ranging from zero to 120 GW for the 50% GHG reduction case and 30-230 gigawatts (GW) for the 80% reduction scenario. While these results indicate the potential importance of CCS as a cost-effective mitigation option for achieving climate goals by mid-century, they also illustrate the large uncertainties in the future demand for CO_2 capture technology and the time frame for its widespread commercial use.

Figure 36. Projected U.S. Energy Mix in 2050 for Two GHG Reduction Scenarios

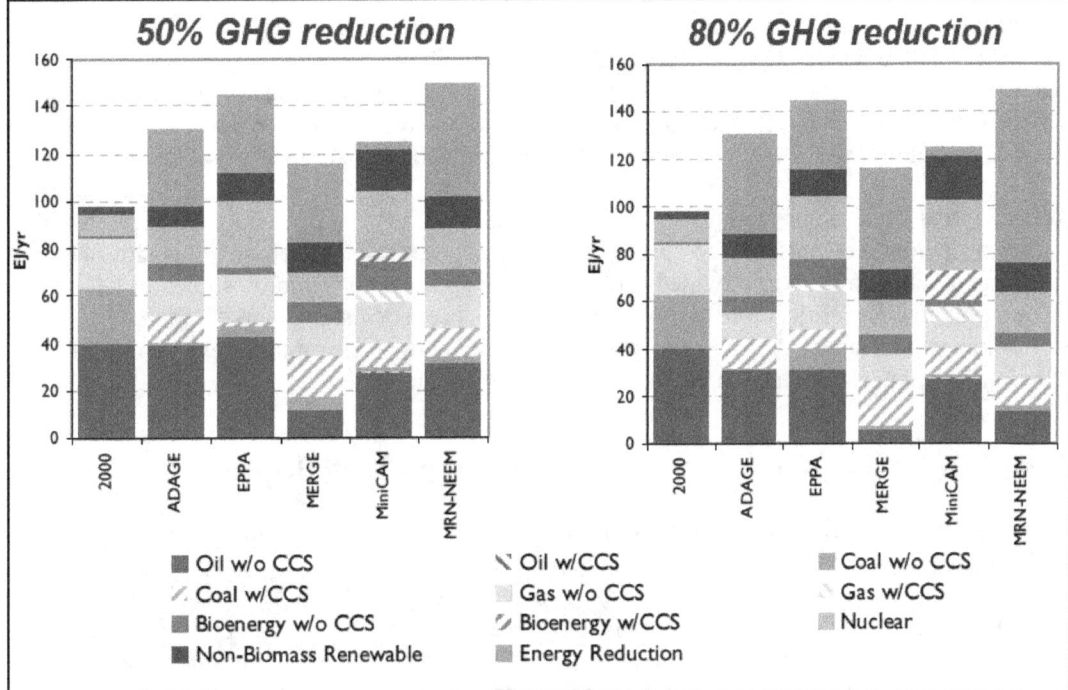

Source: National Research Council, "America's Climate Choices."

Notes: The cross-hatched areas indicate facilities with CCS.

Conclusion

Current roadmaps and scenarios for carbon capture technology commercialization and deployment envision that improved, lower-cost capture systems will be generally available for use at power plants and other industrial facilities by 2020. At the same time, public and private-sector research organizations alike acknowledge that a sustained R&D effort will be required over the next decade to achieve that goal, especially for many of the promising new processes that are still in the early stages of development. The magnitude of future cost reductions also is likely to depend on the pace of CCS technology deployment as well as on continued R&D support. The next chapter looks at past experience with other power plant environmental technologies to provide additional perspectives on the pace of new technology development, deployment, performance improvements, and cost reductions.

---

[102] National Research Council, "America's Climate Choices."

Chapter 9: Lessons from Past Experience

Introduction

This chapter looks retrospectively at a number of other recent efforts to develop and commercialize advanced technologies to improve the effectiveness and lower the cost of air pollutant capture at coal-fired power plants. The purpose of this analysis is to glean insights that are useful for assessing the prospects for improved, lower-cost CO_2 capture systems. First this chapter presents several case studies of prior DOE-supported efforts to develop novel, lower-cost systems to capture power plant sulfur dioxide (SO_2) and nitrogen oxide (NO_x) emissions. These past efforts bear a number of similarities to current efforts for CO_2 capture systems. Thus, they provide some historical benchmarks for the time required to bring a new process from concept to commercialization and the factors that influence the probability of success.

Following this, the chapter presents some historical data on the rates of technology deployment, performance improvements, and cost reductions for post-combustion capture systems of SO_2 and NO_x. Again, the purpose is to provide benchmarks for assessing current projections for CO_2 capture systems. The critical role of government policies in establishing markets for environmental technologies also is discussed and illustrated with examples drawn from past experience with post-combustion SO_2 and NO_x capture technologies.

Case Studies of Novel Capture Technology Development

Current efforts to develop new or improved carbon capture systems are in many respects similar to efforts that began in the late 1970s to develop improved, lower-cost technologies for power plant SO_2 and NO_x controls. Those activities followed passage of the 1970 Clean Air Act Amendments (CAAA) and the adoption of federal New Source Performance Standards (NSPS) requiring "best available control technology" for major new sources of air pollution, including fossil fuel power plants. Although SO_2 capture technology had been used commercially since the early 20th century on various industrial processes (such as metal smelters), it had seldom been used to desulfurize power plant flue gases. The same was true of post-combustion NO_x capture technologies.

By the late 1970s, the most widely used technology for post-combustion SO_2 control (in response to NSPS and CAAA requirements) was a flue gas desulfurization (FGD) system or "scrubber" that used a slurry of water and limestone to capture SO_2 via chemical reactions (analogous to today's CO_2 scrubbers that employ amine-based solutions). These early "wet FGD" systems had high capture efficiencies (up to about 90%), but were widely regarded as being very expensive, being difficult to operate reliably, and having a high energy penalty.[103] In the case of nitrogen oxides, post-combustion capture systems such as selective catalytic reduction (SCR) were deemed too costly and unavailable in the 1970s to be required under the NSPS; instead, a less

---

[103] M. R. Taylor, *The Influence of Government Actions on Innovative Activities in the Development of Environmental Technologies to Control Sulfur Dioxide Emissions from Stationary Sources*, Ph.D. Thesis, Carnegie Mellon University, Pittsburgh, PA, January 2001.

stringent requirement was imposed that did not require post-combustion capture, but instead could be met using only low-NO$_x$ burners.[104]

By the 1980s, U.S. coal-fired power plants were being targeted for further reductions in SO$_2$ and NO$_x$ emissions to curtail the growing problem of acid deposition (acid rain). In response, DOE launched major initiatives to develop "high risk, high payoff" technologies that promised significant cost-effective reductions in power plant SO$_2$ and NO$_x$ emissions compared to the prevailing FGD and SCR technologies.

Five new technologies supported under the DOE Clean Coal Technology program are briefly described below. Three of the novel processes involved post-combustion SO$_2$ and NO$_x$ capture in a single process rather than in separate units. The other two processes sought more cost-effective SO$_2$ capture by injecting solid sorbents directly into the power plant furnace or flue gas duct. Of particular relevance to the present report are the time required to develop each process and its ultimate fate in the commercial marketplace.

The Copper Oxide Process

The use of copper oxide as a sorbent for sulfur removal was first investigated at the laboratory scale by the U.S. Bureau of Mines in 1961.[105] Pilot-scale tests were performed in the mid-1960s and by 1973 the process saw industrial use for sulfur removal at a refinery in Japan.[106] DOE continued to develop the process as a lower-cost way to remove both SO$_2$ and NO$_x$ from power plant flue gases, while producing sulfur or sulfuric acid as a byproduct in lieu of solid waste.[107] **Figure 37** shows several milestones in the process development.

After a series of design changes following pilot plant tests in the 1970s and 1980s, DOE began developing designs for a 500 MW power plant in the 1990s and planned a new 10 MW pilot plant as part of its Low Emission Boiler System project.[108] However, by the time the environmental impact statement for that project was completed, the copper oxide process had been replaced by a conventional wet FGD system.[109] Although the process never developed into a commercial technology for combined SO$_2$ and NO$_x$ capture, research on copper oxide sorbents continues.[110]

---

[104] S. Yeh et al., "Technology Innovations and Experience Curves for Nitrogen Oxides Control Technologies," *Journal of the Air & Waste Management Association*, vol. 55, no. 2 (Dec. 2005), pp. 1827-1838.

[105] D. H. McCrea, A. J. Forney, and J. G. Myers, "Recovery of Sulfur from Flue Gases Using a Copper Oxide Absorbent," *Journal of the Air Pollution Control Association*, vol. 20 (1970), pp. 819-824.

[106] Kohl, "Gas Purification."

[107] H. C. Frey and E. S. Rubin, "Probabilistic Evaluation of Advanced SO$_2$/NO$_x$ Control Technology," *Journal of Air & Waste Management Association*, vol. 41, no. 12 (Dec. 1991), pp. 1585-1593.

[108] U.S. Department of Energy, *Fluidized Bed Copper Oxide Process Phase IV: Conceptual Design and Economic Evaluation*, Report from A. E. Roberts and Associates, Inc., to National Energy Technology Laboratory, Pittsburgh, PA, 1994.

[109] U.S. Department of Energy, "Environmental Impact Statement for the Low Emission Boiler System Proof-of-Concept, Elkbart, Logan County, IL," http://www.netl.doe.gov/technologies/coalpower/cctc/cctdp/bibliography/misc/pdfs/hipps/000001B3.pdf.

[110] J. Abbasian and V. S. Gavaskar, "Dry Regenerable Metal Oxide Sorbents for SO$_2$ Removal from Flue Gases. 2. Modeling of the Sulfation Reaction Involving Copper Oxide Sorbents," *Industrial & Engineering Chemical Research*, vol. 46, no. 4 (2007), pp. 1161-1166. T. Benko and P. Mizsey, "Comparison of Flue Gas Desulfurization Processes Based On Lifecycle Assessment," *Chemical Engineering*, vol. 51, no. 2 (2007), pp. 19-27.

Figure 37. Development History of the Copper Oxide Process for Post-Combustion SO₂ and NOₓ Capture

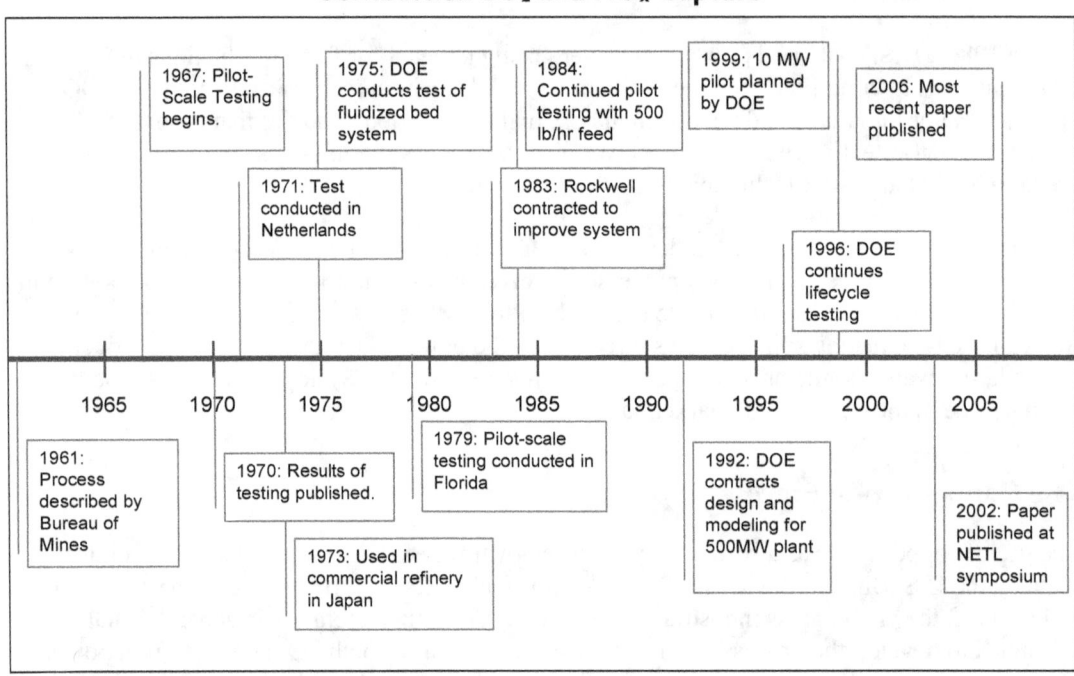

Source: Edward S. Rubin, Aaron Marks, Hari Mantripragada, Peter Versteeg, and John Kitchin, Carnegie Mellon University, Department of Engineering and Public Policy.

The Electron Beam Process

The electron beam process for flue gas treatment was first introduced by the Ebara Corporation of Japan in 1970.[111] The concept was that energy from the electron beam would excite chemicals in the flue gas, causing them to break down and form other stable compounds. The process was promoted as a more cost-effective way to simultaneously capture both SO_2 and NO_x with high (~90%) efficiency. **Figure 38** shows the history of key process developments.

By 1977, Ebara's testing moved to the pilot plant scale and in 1985 their subsidiaries in the United States and Germany opened two more pilot plants, one in Indiana and one in Germany.[112] DOE provided partial funding for the U.S. facility. Continued R&D led to the first commercial plant in China in 1998, followed by three more plants built between 1999 and 2005, one in Poland, the other two in China.[113] The overall cost of this system is highly dependent on the market value of the ammonium sulfate and ammonium nitrate byproducts that are produced, as well as on the cost of ammonia, the key reagent for the process. The need for these byproduct chemicals may help explain the adoption of this process in China. There have been no commercial installations of the electron beam process in the United States.

---

[111] Kohl, "Gas Purification."

[112] V. Markovic, "Electron Beam Processing of Combustion Flue Gases," IAEA Bulletin, No. 3, International Atomic Energy Agency, 1987, Vienna, Austria.

[113] I. Calinescu et al., "Electron Beam Technologies for Reducing SO_2 and NO_x Emissions from Thermal Power Plants," Proc. *World Energy Council Regional Energy Forum, 2008*, Neptun, Romania.

Figure 38. Development History of the Electron Beam Process for Post-Combustion SO₂ and NOₓ Capture

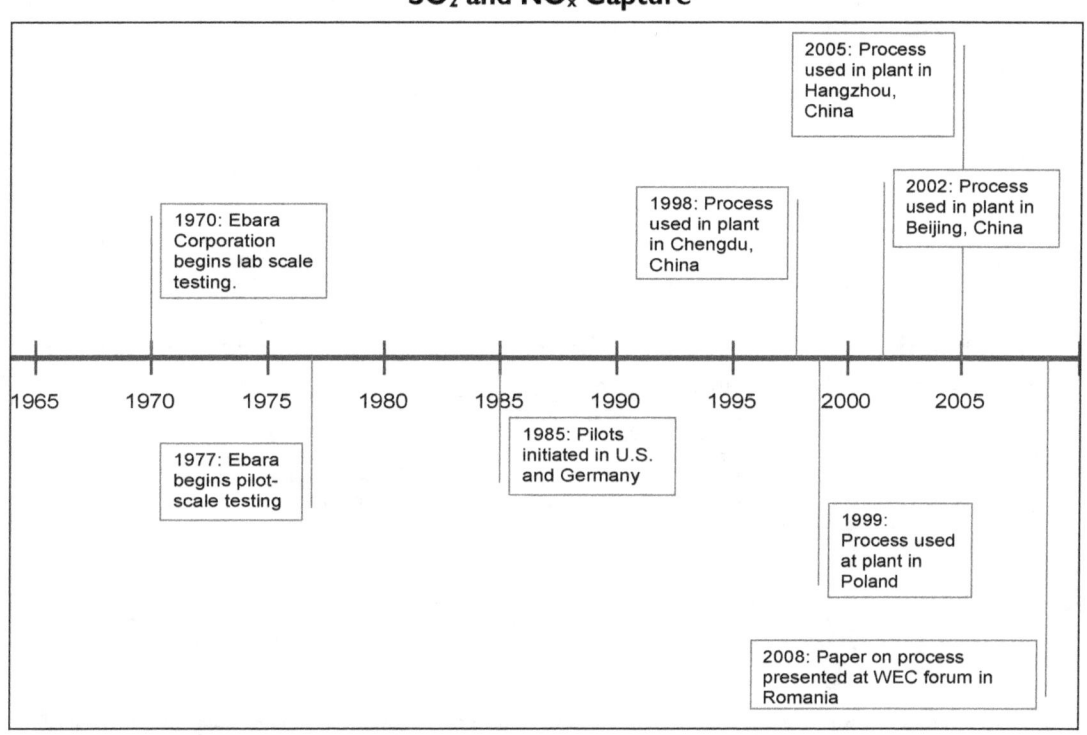

Source: Edward S. Rubin, Aaron Marks, Hari Mantripragada, Peter Versteeg, and John Kitchin, Carnegie Mellon University, Department of Engineering and Public Policy.

The NOXSO Process

The NOXSO process was another concept for post-combustion capture of both SO_2 and NO_x from power plant flue gases. It used a solid sorbent of sodium carbonate supported on alumina beads. The sorbent chemistry was based on an alkalized alumina process developed by the US Bureau of Mines in the 1960s. A novel feature the NOXSO process was the use of a fluidized bed reactor for sorbent regeneration. **Figure 39** shows the process development timeline, which began in 1979 with funding from DOE.

Pilot plant and life cycle testing were carried out from 1982 to 1993. In 1991 the NOXSO Corporation received a DOE contract to build a commercial-scale demonstration plant.[114] However, a number of administrative problems ensued, leading to several changes in the project site location. A legal dispute with the owner of the final project site culminated in the bankruptcy and subsequent liquidation of the NOXSO Corporation.[115]

---

[114] U.S. Department of Energy, "Comprehensive Report to Congress: Commercial Demonstration of the NOXSO SO_2/NO_x Removal Flue Gas Cleanup System," 1991, Washington, DC.

[115] *Chemical Week*, "NOXSO Sues Olin Over SO_2 Agreement," 1997, vol. 159.

Figure 39. Development History of the NOXSO Process for Post-Combustion SO₂ and NOₓ Capture

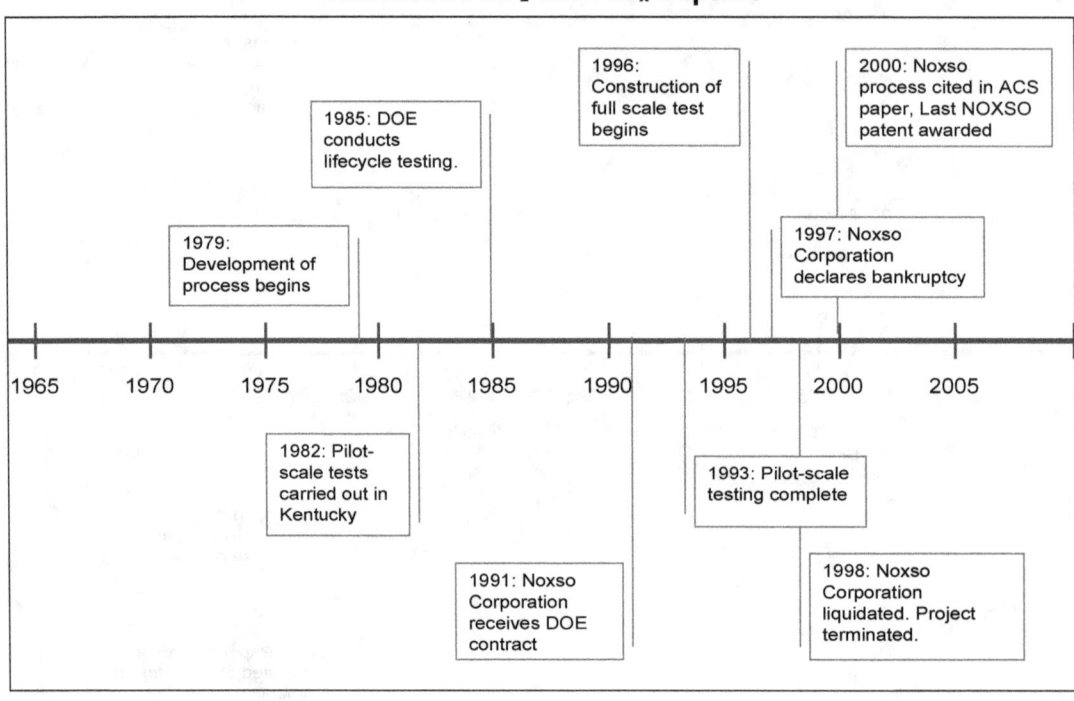

Source: Edward S. Rubin, Aaron Marks, Hari Mantripragada, Peter Versteeg, and John Kitchin, Carnegie Mellon University, Department of Engineering and Public Policy.

The Furnace Limestone Injection Process

In the early 1980s, the prospect of new restrictions on SO_2 emissions to control acid rain prompted interest in sulfur removal methods that were more cost-effective than FGD (post-combustion capture) systems, especially for existing power plants. The furnace limestone injection process promised to be such a technology. Limestone sorbent would be injected directly into the furnace and react with sulfur oxides to achieve only moderate removal efficiencies, but at very low cost. The method was first tested by Wisconsin Power in 1967.[116] In the 1980s and 1990s, DOE supported two methods of furnace sorbent injection (called LIFAC and LIMB), as seen in **Figure 40**.

The LIFAC process combined limestone injection with a humidification system to capture SO_2. First developed by the Tampella Company in 1983, it was later tested at a commercial scale in Finland. DOE supported demonstrations in the United States starting in 1990, achieving 70% to 80% sulfur removal rates.[117] The LIMB (limestone injection with multi-stage burners) process was first developed by the U.S. Environmental Protection Agency. It achieved fairly low (approximately 50%) SO_2 removal using limestone, with somewhat higher capture efficiencies using more expensive lime sorbents. Testing of both processes encountered failures of the electrostatic precipitator at the test plants due to the larger volume of solids being collected.

---

[116] W. A. Pollock et al., *Mechanical Engineering*, American Society of Mechanical Engineers, New York, NY, 1967.

[117] U.S. Department of Energy, *LIFAC Sorbent Injection Desulfurization Demonstration Project: A DOE Assessment*, National Energy Technology Laboratory, Pittsburgh, PA, 2001.

Technical solutions added to the cost. [118] The LIFAC process was eventually used commercially at nine facilities outside the United States, but neither LIFAC nor LIMB was adopted commercially for SO_2 control in the United States following the large-scale demonstrations.

Figure 40. Development History of the Furnace Limestone Injection Process for SO_2 Capture

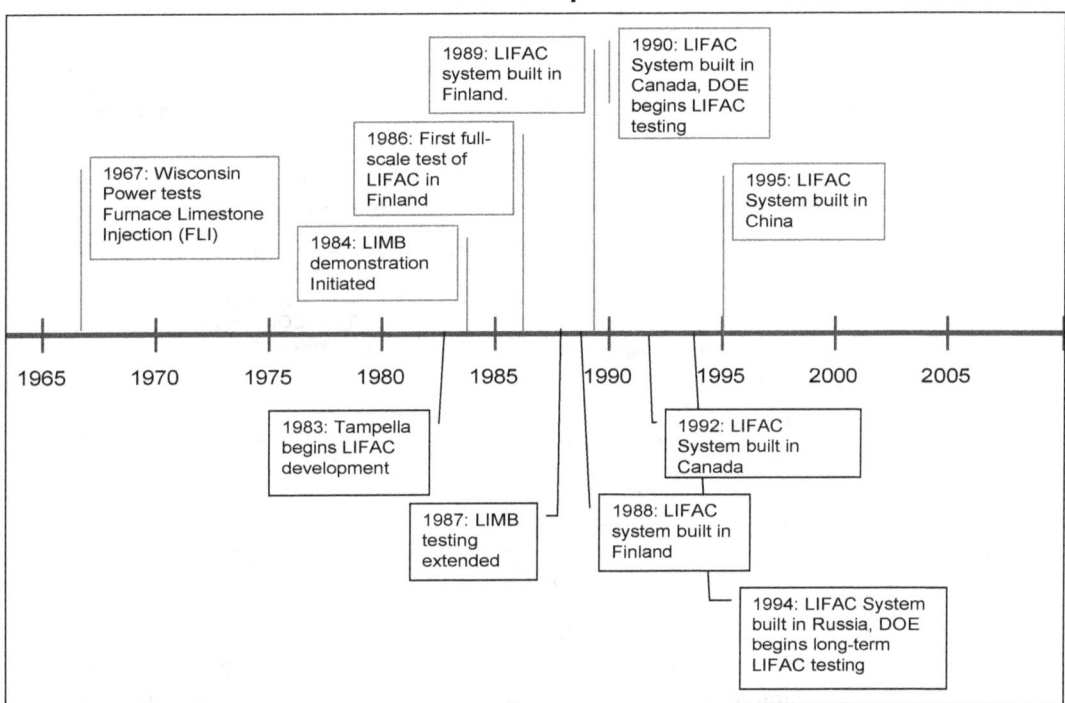

Source: Edward S. Rubin, Aaron Marks, Hari Mantripragada, Peter Versteeg, and John Kitchin, Carnegie Mellon University, Department of Engineering and Public Policy.

The Duct Sorbent Injection Process

Duct sorbent injection (DSI) is another post-combustion SO_2 capture concept similar to furnace limestone injection, except that the sorbent is injected into the flue gas duct after the boiler where temperatures are lower and physical access is generally easier. This was proposed as a simpler and more cost-effective method of achieving modest SO_2 reductions at existing power plants. **Figure 41** shows the process development timeline.

---

[118] Southern Research Institute, *Analysis of Whitewater Valley Unit 2 ESP Problems during Operation of the LIFAC SO_2 Control Process*, Report SRI-ENV-93-953-7945-I, Prepared for Southern Company Services by Southern Research Institute, Birmingham, AL, August 1993. Babcock and Wilcox, *LIMB Demonstration Project Extension and Coolside Demonstration*, Report No. DOE/PC/7979-T27 to the U.S. Department of Energy, Pittsburgh Energy Technology Center, Prepared by T.R Goots et al., The Babcock & Wilcox Company, Barberton, OH, October 1992.

Figure 41. Development History of the Duct Sorbent Injection Process for SO₂ Capture

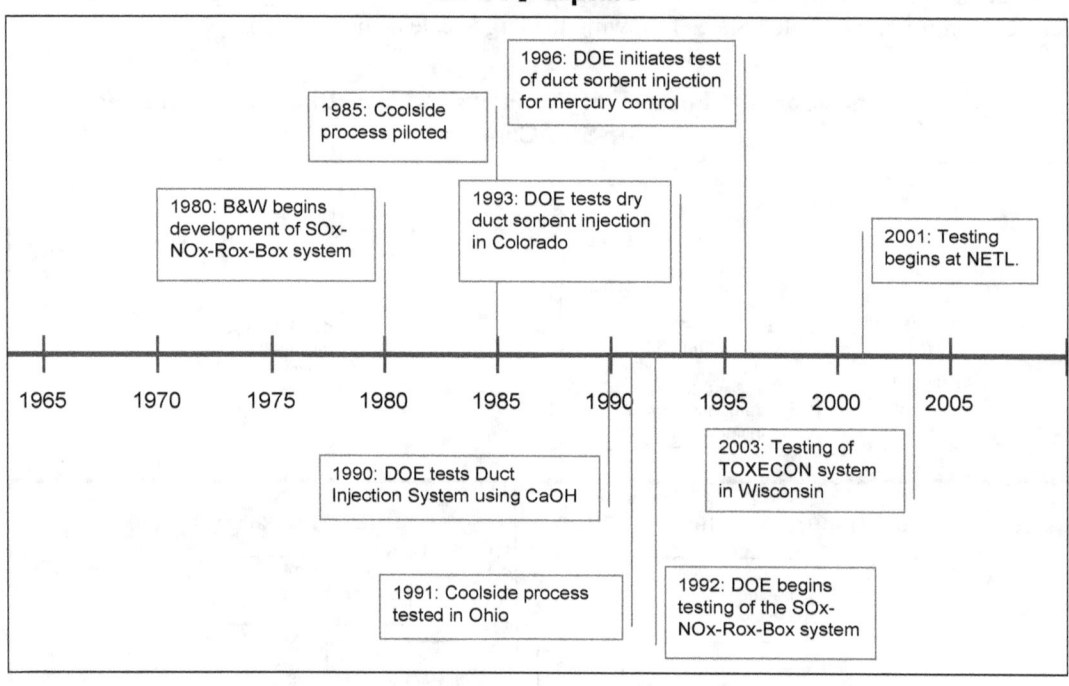

Source: Edward S. Rubin, Aaron Marks, Hari Mantripragada, Peter Versteeg, and John Kitchin, Carnegie Mellon University, Department of Engineering and Public Policy.

Babcock and Wilcox began work on a DSI system in 1980 for their SOx-NOx-ROx-BOx (SNRB) combined pollutant control system, which DOE tested 12 years later. Pilot and demonstration projects of DSI for SO₂ capture during the 1980s and early 1990s achieved capture rates rarely exceeding 40% with calcium-based sorbents. Costs and technical complexity were similar to the more effective furnace injection systems.[119] Subsequent process modifications improved the SO₂ capture efficiency, but at a higher cost. There were no commercial adoptions of DSI following the DOE test programs.

In 1996, DSI was retooled for use in mercury control. It developed into the TOXECON process, which was tested at full scale in 2003, achieving 90% capture of flue gas mercury.[120] Duct sorbent injection for mercury control is now offered commercially but has not been widely adopted in light of continuing uncertainty over final national power plant mercury emissions regulations.

Implications for Advanced Carbon Capture Systems

Several lessons can be gleaned from the case studies above that are relevant to current efforts to develop lower-cost carbon capture systems for power plants. The first is the importance of markets for these environmental technologies. Just as with advanced CO₂ capture systems today,

---

[119] T. Hunt et al., "Performance of the Integrated Dry NOx/SO₂ Emissions Control System," U.S. Department of Energy, Fourth Annual Clean Coal Technology Conference, September 1995, Denver, CO.

[120] ADA Environmental Solutions, *TOXECON Retrofit for Multi-Pollutant Control on Three 90-MW Coal Fired Boilers*, Report to U.S. Department of Energy, National Energy Technology Laboratory, prepared by ADA Environmental Solutions (2008).

at the time they were being developed there were no requirements for (hence, no significant markets for) high-efficiency combined SO_2-NO_x capture systems, or moderately efficient SO_2 removal systems. This factor alone posed high risks for their commercial success. While that was consistent with the DOE mission to pursue high-risk, high-payoff technologies, the high payoffs that were projected never materialized—in large part because the markets for these technologies failed to develop as expected. Similar risks face advanced carbon capture technologies today.

As shown in the figures above, the time required to develop a novel capture process from concept to large-scale demonstration was typically two decades or more. During this period the projected economic benefits of the advanced technologies tended to shrink. Not only did their cost tend to rise during the development process (as suggested earlier in **Figure 26**), but the cost of competing options also fell. Thus, the continual deployment and improvement of commercial FGD systems (mainly in the United States) and SCR systems (in Japan and Germany) during the 1980s made it increasingly difficult for combined SO_2-NO_x capture technologies to enter and compete in the marketplace. Indeed, in the United States, there was no market for post-combustion NO_x capture at coal-burning plants until the mid-1990s.[121] In the case of furnace and duct sorbent injection processes for moderate levels of SO_2 capture, the anticipated market for such an option did materialize in the United States with passage of the acid rain provisions of the 1990 Clean Air Act Amendments. However, switching to low-sulfur coal proved to be an easier and more economical choice than sorbent injection, especially as low-sulfur western coals entered the marketplace.

In terms of additional lessons learned, the above discussion suggests that the lengthy time historically required to develop advanced environmental technologies tends to diminish the probability of commercial success, as more mature technologies gain initial market share (assuming the existence of a market). Thus, any efforts that can accelerate the development process can help reduce the commercial risks. Apropos of that, another lesson drawn from this experience is that current commercial technologies do not "stand still" (as is often assumed by proponents of new technologies). Advancements in current systems also must be anticipated to more realistically assess the prospects and potential payoffs of advanced technologies that are still under development.

The Pace of Capture Technology Deployment

Historical rates of deployment for other power plant environmental technologies can serve as a useful guide for realistically assessing current projections for CCS technologies.

Figure 42 shows the trends in deployment of post-combustion capture systems for SO_2 and NO_x from 1970 to 2000. For FGD systems, the maximum rate of deployment in response to new environmental policy requirements over this period was approximately 15 GW per year (in Germany), with an average rate of about 8 GW per year worldwide. For SCR systems, the maximum rate was about 10 GW per year (again in Germany), with an average global deployment rate of about 5 GW per year. These results suggests that deployment scenarios for CO_2 capture systems that significantly exceed these rate may be unrealistic or will require aggressive new efforts and measures to achieve.

---

[121] Yeh et al., "Technology Innovations."

Figure 42. Historical Deployment Trends for Post-Combustion SO₂ and NOₓ Capture Systems (FGD and SCR Technologies)

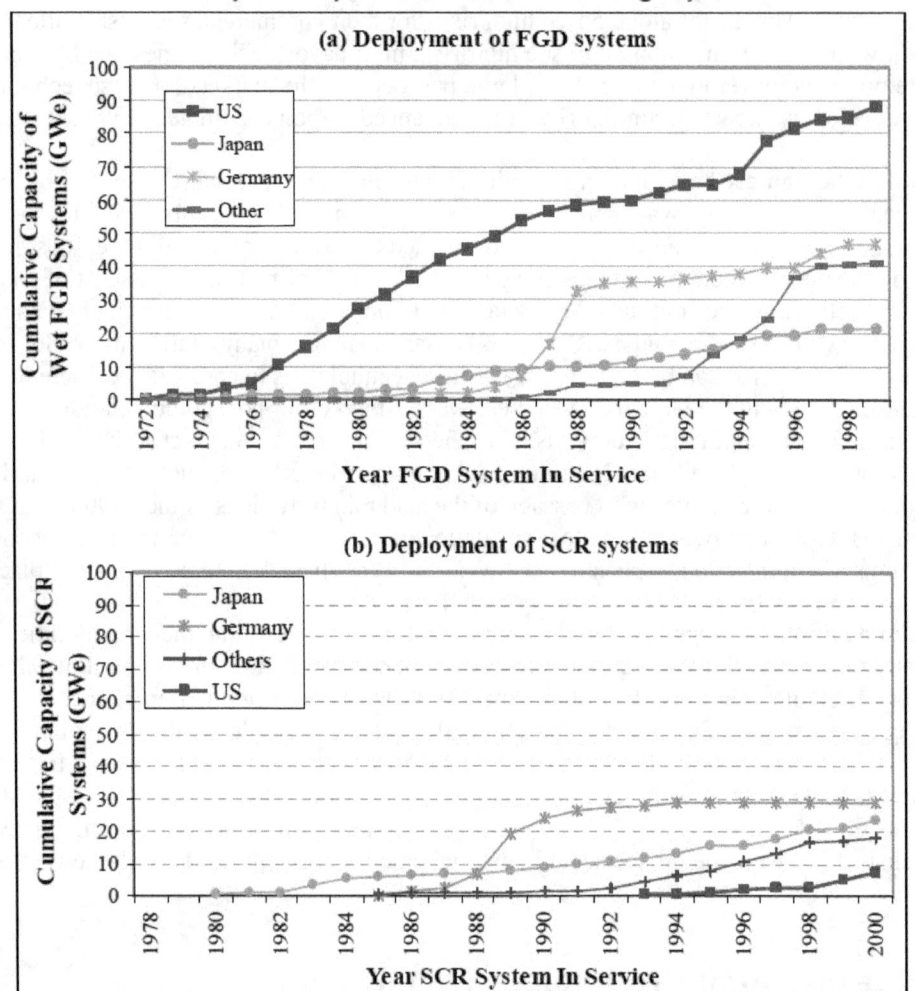

Source: E. S. Rubin et al., "Use of Experience Curves to Estimate the Future Cost of Power Plants with CO₂ Capture," *International Journal of Greenhouse Gas Control*, vol. 1, no. 2 (2007), pp. 188-197.

Rates of Performance and Cost Improvements

Studies also have documented the historical rates of improvement in the performance (capture efficiency) of power plant emission control systems and their rates of cost reduction following commercialization.[122] For example, **Figure 43** shows the trend in average SO₂ capture efficiency for power plant FGD systems coming online from 1969 to 1995. Capture efficiencies increased from about 70% to 95% over that period due to the combined effects of technology improvements and regulatory requirements. Since that time the performance of wet FGD systems has continued to improve, with new systems today capturing 98% to 99% or more of the SO₂. These deep levels of sulfur removal now can facilitate post-combustion CO₂ capture systems, which require inlet

---

[122] J. Longwell, E. S. Rubin, and J. Wilson, "Coal: Energy for the Future," *Progress in Energy and Combustion Science*, vol. 21 (1995), pp. 269-360; Rubin et al., "Use of Experience Curves."

SO_2 concentrations as low as one part per million for some commercial amine-based systems.[123] These data, as well as other historical trends of increasing capture efficiencies for power plant particulates and SO_2 and NO_x emissions,[124] suggest the potential for future improvements in commercial CO_2 capture systems as well.

Figure 43. Improvements in SO₂ Removal Efficiency of Commercial Lime and Limestone FGD Systems Coming Online in a Given Year, as a Function of Cumulative Installed FGD Capacity in the United States

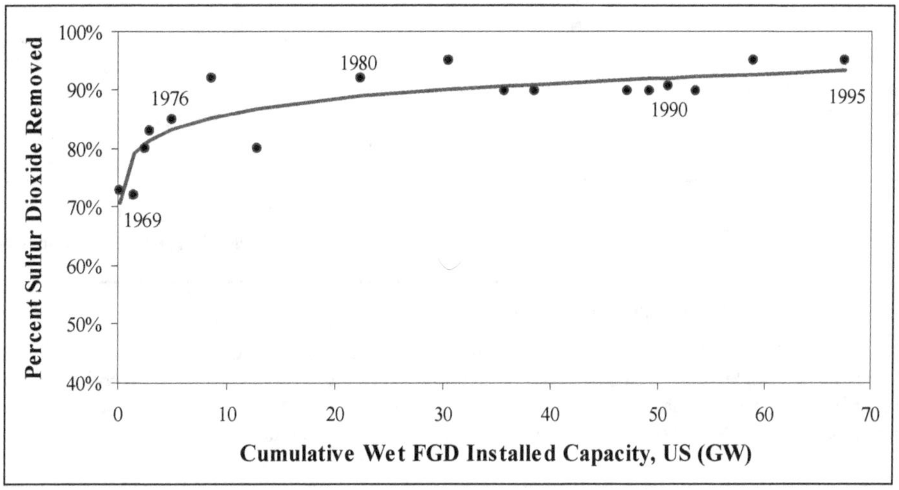

Source: E. S. Rubin et al., *The Effect of Government Actions on Environmental Technology Innovation: Applications to the Integrated Assessment of Carbon Sequestration Technologies*, report from Carnegie Mellon University, Pittsburgh, PA, to U.S. Department of Energy, Germantown, MD, January 2004, p. 153.

Figure 44 shows the historical trends in capital costs for FGD and SCR systems on standardized coal-fired power plants in the United States. In both cases, the actual or estimated capital cost (as well as O&M costs) increased during the early commercialization of these technologies in order to achieve the levels of availability and performance required for utility operations. Subsequently, costs declined considerably with increasing deployment. On average, the capital cost of these technologies fell by 13% for each doubling of total installed capacity.[125] This "learning rate" was also assumed for future CO_2 capture systems in the plant-level cost projections shown earlier in **Figure 29**.

---

[123] Mitsubishi Heavy Industries, "KM-CDR Post-Combustion CO_2 Capture with KS-1 Advanced Solvent," Eighth Annual Conference on Carbon Capture and Sequestration, May 4-7, 2009, Pittsburgh, PA, Exchange Monitor Publications, Washington, DC.

[124] Longwell et al., "Coal: Energy."

[125] Rubin et al., "Cost and Performance."

Figure 44. Capital Cost Trends for Post-Combustion Capture of SO₂ and NOₓ at a New Coal-Fired Power Plant

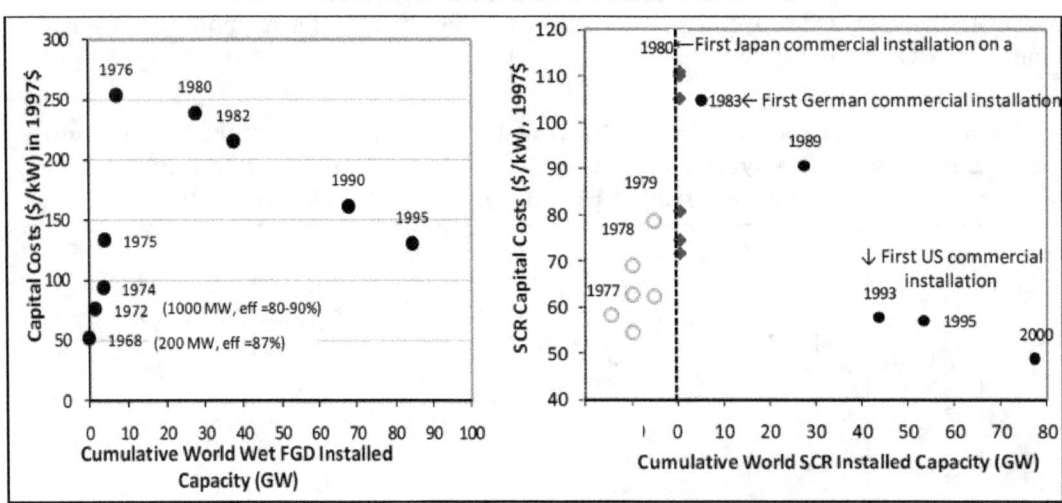

Source: Rubin et al., "Use of Experience Curves."

Notes: On the left, capital cost trend for a wet limestone FGD system at a standardized new power plant (500 MW, 3.5% sulfur coal, 90% SO₂ removal, except where noted); on the right, capital cost trend for a SCR system at a new plant (500 MW, medium sulfur coal, 80% NOₓ removal). Solid diamond symbols are studies based on low-sulfur coal plants. Open circles are studies prior to SCR use on coal-fired power plants.

The Critical Role of Government Actions

In the U.S. economy, the existence of a market (or demand) for a product is critical to its adoption and widespread use. This is true as well for CO_2 capture technologies. The adoption and diffusion of a technology also are key elements of the innovation process that improves performance of a product and reduces its cost over time, as depicted earlier in **Figure 10**. R&D plays a critical role in this process. But R&D alone is not sufficient without a market for the technology.

For environmental technologies such as CO_2 capture and storage systems, few if any markets exist in the absence of government actions and policies. What electric utility company, for example, would want to spend a large sum of money to install CCS—even with an improved lower-cost capture process—if there is no requirement or incentive to reduce CO_2 emissions? A costly action such as this provides little or no economic value to the company—indeed, the added cost and energy penalty of CO_2 capture increase the cost of operation. Only if a government action either required CO_2 capture and storage, or made it financially worthwhile to reduce CO_2 emissions, would a sizeable market be created for technologies that enable such reductions.

Thus, as with other environmental emissions that affect the public welfare, government actions are needed to create or enhance markets for CO_2 emission-reducing technologies.

Different policy measures influence markets in different ways. Measures such as government loan guarantees, tax credits, direct financial subsides, and R&D funding can help create markets by providing incentives for technology development, deployment, and diffusion. Voluntary incentives such as these are commonly referred to as "technology policy" measures.[126] In contrast, regulatory policies such as an emissions cap, emissions tax, or performance standards that limit emissions to specified levels are mandatory, not voluntary. These policies create or expand markets for lower-emission technologies by imposing requirements that can be met only—or most economically—by the use of a low-emission technology.

Through their influence on markets for environmental technologies (like CO_2 capture and storage systems), government actions also are a critical element of the technological innovation process. Studies of past measures to reduce sulfur dioxide and nitrogen oxide emissions from U.S. power plants have documented the ability of regulatory policies to influence both the magnitude and direction of efforts to develop new or improved capture technologies.[127] **Figure 45** and **Figure 46**, for example, show the century-scale trends in U.S. patenting activity for SO_2 reduction technologies and post-combustion NO_x capture systems, respectively. In both cases, the number of new patents filed—a measure of "inventive activity"—increased dramatically when new environmental regulations that required or incentivized the use of these technologies came into force. (In the case of NO_x control, such regulations for coal plants came first in Japan and Germany; U.S. regulations lagged by more than a decade.) The subsequent reduction in cost that accompanied the increased deployment of these technologies (**Figure 46**) is evidence of the influence of government actions on technology innovations in this domain.

Figure 45. Trend in U.S. Patenting Activity for SO_2 Removal Technologies

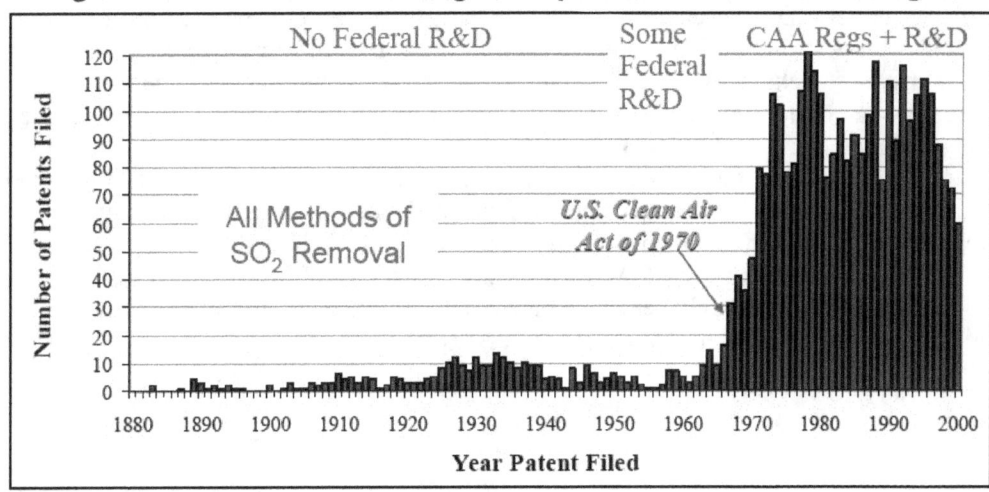

Source: Adapted from Taylor et al., "Control of SO_2."

---

[126] J. A. Alic, D. S. Mowery, and E. S. Rubin, *U.S. Technology and Innovation Policies: Lessons for Climate Change*, Pew Center on Global Climate Change, 2003.

[127] M. R. Taylor, E. S. Rubin, and D. A. Hounshell, "Control of SO_2 Emissions from Power Plants: A Case of Induced Technological Innovation in the U.S.," *Technological Forecasting and Social Change*, vol. 72, no. 6 (July 2005), pp. 697-718. Yeh, "Technology Innovations."

Figure 46. Trend in U.S. Patenting Activity for Post-Combustion NO$_x$ Removal Technologies

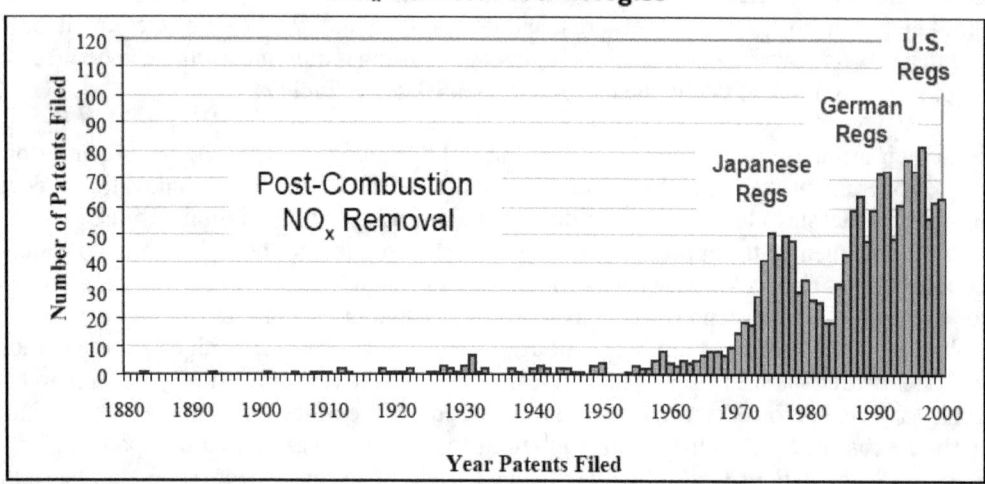

Source: Yeh, et al., "Technology Innovations."

Conclusion

This chapter has examined recent historical experience in the development of advanced technologies for post-combustion capture of sulfur dioxide and nitrogen oxide emissions at coal-fired power plants, seeking lessons and insights relevant to current programs to develop improved, lower-cost technologies for CO$_2$ capture. The analysis revealed that several decades were commonly required to develop a new process from concept to a commercial-scale demonstration. It also illustrated the risks inherent in developing new environmental technologies for which there is not yet a significant market. Benchmark rates of capture technology deployment and long-term cost reductions also were derived from United States and global experience with FGD systems (for SO$_2$ capture) and SCR systems (for NO$_x$ capture). These historical data underscore the challenging nature of current plans and roadmaps for the commercialization of advanced CO$_2$ capture processes.

Chapter 10: Discussion and Conclusions

This report has sought to provide a realistic assessment of prospects for improved, lower-cost CO_2 capture systems for use at power plants and other industrial facilities in order to mitigate emissions of greenhouse gases linked to global climate change. Toward that end, the report first described each of the three current approaches to CO_2 capture, namely, post-combustion capture from power plant flue gases using amine-based solvents such as monoethanolamine (MEA); pre-combustion capture (also via chemical solvents) from the synthesis gas produced in an integrated coal gasification combined cycle power plant; and oxy-combustion capture, in which high-purity oxygen is used for combustion to produce a flue gas with high CO_2 concentration amenable to capture without a post-combustion chemical process.

Currently, post-combustion and pre-combustion capture technologies are commercial and widely used for gas stream purification in a variety of industrial processes, including several small-scale installations on power plant flue gases that produce commodity CO_2 for sale. Oxy-combustion capture is still under development and is not currently commercial. The advantages and limitations of each of these three methods are discussed in this report, along with plans for their continued development and demonstration in large-scale power plant applications.

While all three approaches are capable of high CO_2 capture efficiencies (typically about 90%), major drawbacks of current processes are their high cost and large energy requirements for operation (which contribute significantly to the high cost). This is especially true for the combustion-based capture processes, which have the highest incremental cost relative to a similar plant without CO_2 capture.

Also discussed in this report are the substantial R&D activities underway in the United States and elsewhere to develop and commercialize improved solvents that can lower the cost of current post-combustion capture processes, as well as research on a variety of potential "breakthrough technologies" such as novel solvents, sorbents, membranes, and oxyfuel systems that hold promise for lower-cost capture systems. Most of these processes, however, are still in the early stages of research and development (i.e., conceptual designs and laboratory- or bench-scale processes), so that credible estimates of their performance and (especially) cost are lacking at this time. Even with an aggressive development schedule, the commercial availability of these technologies, should they prove successful, is at least a decade away based on past experience.

Processes at the more advanced pilot plant scale are, for the most part, new or improved solvent formulations (such as ammonia and advanced amines) that are undergoing testing and evaluation. These advanced solvents could be available for commercial use within several years if subsequent full-scale testing confirms their overall benefit. Pilot-scale oxy-combustion processes also are currently being tested and evaluated for planned scale-up, while in Europe two IGCC plants are installing pilot plants to evaluate pre-combustion capture options.

At the moment, however, there are still no full-scale applications of CO_2 capture at a coal-based power plant, although a number of demonstration projects are planned or underway in the United States and other countries. Capture projects for other types of industrial facilities also are planned.

In general, the focus of most current R&D activities is on cost reduction rather than additional gains in the efficiency of CO_2 capture (which can often result in higher overall cost). While a number of programs emphasize the need for lower-cost retrofit technologies suitable for existing power plants, as a practical matter these same technologies are being pursued to reduce capture costs for new plant applications as well. Indeed, as the fleet of existing coal-fired power plants continues to age, the size of the potential U.S retrofit market for CO_2 capture will continue to shrink, as older plants may not be economic to retrofit (although the situation in other countries, especially China, may be quite different).

Whether for new power plants or existing ones, the key questions are, when will advanced CO_2 capture systems be available for commercial rollout, and how much cheaper will they be compared to current technology?

All of the technology roadmaps reviewed in this report anticipate that CO_2 capture will be available for commercial deployment at power plants by 2020. For current commercial technologies like post-combustion amine systems, this is a conservative estimate, since the key requirement is for scale-up and demonstration at a full-size power plant—achievable well before 2020. A number of roadmaps also project that novel, lower-cost technologies like solid sorbent systems for post-combustion capture also will be commercial in the 2020 time frame. Such projections acknowledge, however, that this will require aggressive and sustained efforts to advance promising concepts to commercial reality.

That caveat is strongly supported by our review of recent experience from R&D programs to develop lower-cost technologies for post-combustion SO_2 and NO_x capture at coal-fired power plants. Those efforts typically took two decades or more to bring a new concept (like combined SO_2 and NO_x capture systems) to commercial availability. By then, the cost advantage initially foreseen had largely evaporated: advanced technologies tended to get more expensive as the development process progressed (consistent with "textbook" descriptions of the innovation process), while the cost of formerly "high-cost" commercial options gradually declined over time. In a number of cases, the absence of a market for the advanced technology (as is currently the case for CO_2 capture systems) put it at a further disadvantage.

The good news based on past experience is that the costs of environmental technologies that succeed in the marketplace tend to fall over time. For example, after an initial rise during the early commercialization period, the cost of post-combustion SO_2 and NO_x capture systems declined by 50% or more after about two decades of deployment at coal-fired power plants. This trend is consistent with the "learning curve" behavior seen for many other classes of technology. It thus appears reasonable to expect a similar trend for future CO_2 capture costs once these technologies become widely deployed. This report also notes that the cost of CO_2 capture also depends strongly on other aspects of power plant design, financing, and operation—not solely on the cost of the CO_2 capture unit. Future improvements in net power plant efficiency, for example, will tend to lower the unit cost of CO_2 capture.

Some estimates of future electricity generation costs for advanced power plant designs with CO_2 capture and storage offer even more optimistic forecasts of potential cost reductions from advanced technologies. In general, however, the further away a technology is from commercial reality, the lower its estimated cost. Thus, there is considerable uncertainty in the projected cost of technologies that are not yet commercial, especially those that exist only as conceptual designs.

More reliable estimates of future technology costs typically are linked to projections of their expected level of commercial deployment in a given time frame (i.e., a measure of their market size). For power plant technologies like CO_2 capture systems, this is commonly expressed as total installed capacity. However, as with other technologies whose sole purpose is to control environmental emissions, there is no significant market for power plant CO_2 capture systems absent government actions or policies that effectively create such markets—either through regulations that limit CO_2 emissions or through voluntary incentives for its use. The historical evidence and technical literature examined in this report strongly link future cost reductions to the level of commercial deployment of a technology. In empirical "experience curve" models, the latter measure serves as a surrogate for the many factors that influence future costs, including expenditures for R&D and the knowledge gained through learning-by-doing (related to manufacturing) and learning-by-using (related to technology use).

Based on such models, published estimates project the future cost of electricity from power plants with CO_2 capture to fall by up to 30% below current values after roughly 100,000 MW of capture plant capacity has been installed and operated worldwide. That would represent a significant decrease from current costs—one that would bring the cost and efficiency of future power plants with CO_2 capture close to that of current plants without capture. For reference, it took approximately 20 years following passage of the 1970 Clean Air Act Amendments to achieve a comparable level of technology deployment for SO_2 capture systems at coal-fired power plants.

Uncertainty estimates for these projections, however, indicate that future cost reductions for CO_2 capture also could be much smaller than indicated above. Thus, whether future cost reductions will meet, exceed, or fall short of current estimates will only be known with hindsight.

In the context of this report, the key insight governing prospects for improved carbon capture technology is that achieving significant cost reductions will require not only a vigorous and sustained level of R&D, but also a substantial level of commercial deployment. That will require a significant market for CO_2 capture technologies, which can only be established by government actions. At present such a market does not yet exist. While various types of incentive programs can accelerate the development and deployment of CO_2 capture technology, actions that significantly limit emissions of CO_2 to the atmosphere ultimately are needed to realize substantial and sustained reductions in the future cost of CO_2 capture.

Author Contact Information

Peter Folger
Specialist in Energy and Natural Resources Policy
pfolger@crs.loc.gov, 7-1517

www.ingramcontent.com/pod-product-compliance
Lightning Source LLC
Chambersburg PA
CBHW081828170526
45167CB00007B/2750